Exploring Linear Algebra
Labs and Projects with MATLAB®

Textbooks in Mathematics

Series editors:
Al Boggess and Ken Rosen

Exploring Linear Algebra
Labs and Projects with MATLAB®

Crista Arangala

CRC Press
Taylor & Francis Group
Boca Raton London New York

CRC Press is an imprint of the
Taylor & Francis Group, an **informa** business

CRC Press
Taylor & Francis Group
6000 Broken Sound Parkway NW, Suite 300
Boca Raton, FL 33487-2742

Printed on acid-free paper
Version Date: 20190117

International Standard Book Number-13: 978-1-138-06351-8 (Hardback)
International Standard Book Number-13: 978-1-138-06349-5 (Paperback)

Library of Congress Cataloging-in-Publication Data

Names: Arangala, Crista, author.
Title: Exploring linear algebra : labs and projects with Matlab / Crista Arangala.
Description: Boca Raton : CRC Press, Taylor & Francis Group, 2019. | Includes bibliographical references and index.
Identifiers: LCCN 2018054578 | ISBN 9781138063495
Subjects: LCSH: Algebras, Linear--Computer-assisted instruction. | MATLAB.
Classification: LCC QA185.C65 A73 2019 | DDC 512/.5028553--dc23
LC record available at https://lccn.loc.gov/2018054578

Visit the Taylor & Francis Web site at
http://www.taylorandfrancis.com

and the CRC Press Web site at
http://www.crcpress.com

Contents

Preface

This text is meant to be a hands-on lab manual that can be used in class every day to guide the exploration of linear algebra. Most lab exercises consist of two separate sections, explanations of material with integrated exercises, and theorems and problems.

The exercise sections integrate problems, technology (MATLAB R2017b), MATLAB visualization, and MATLAB simulations that allow students to discover the theory and applications of linear algebra in a meaningful and memorable way. It is important to note that on a very few occasions, the Symbolize Toolbox features that are included in MATLAB R2017b, and not in previous versions, are implemented.

The intention of the theorems and problems section is to integrate the theoretical aspects of linear algebra into the classroom. Instructors are encouraged to have students discover the truth of each of the theorems and proofs, to help their students move toward proving (or disproving) each statement, and to allow class time for students to present their results to their peers. If this course is also serving as an introduction to proofs, we encourage the professor to introduce proof techniques early on as the theorem and problems sections begin in Lab 3.

There are a total of 80 theorems and problems introduced throughout the labs. The author has intentionally labeled those results that are traditional linear algebra theorems as theorems in these sections and has labeled other significant results and interesting problems as problems. There are, of course, many more results, and users are encouraged to make conjectures followed by proofs throughout the course.

In addition, each chapter contains a project set that consists of application-driven projects that emphasize the material in the chapter. Some of these projects are extended in follow-up chapters, and students should be encouraged to use many of these projects as the basis for further undergraduate research.

MATLAB ® is a registered trademark of The MathWorks, Inc. For product information please contact:
The MathWorks, Inc.
3 Apple Hill Drive
Natick, MA, 01760-2098 USA
Tel: 508-647-7000
Fax: 508-647-7001
E-mail: info@mathworks.com
Web: www.mathworks.com

Acknowledgments

Each time I publish a book, my father, Joseph Coles, jokingly asks if I have dedicated the book to him. I have made dedications to my children, to my colleagues, and to my students, but I really would never have gotten to where I am today if my parents, Joseph and Carol Ann Coles, had not taught me to be strong and confident. So this one is for you Dad and Mom. Thanks for all your support.

The writing of this text was supported by an Elon University Funds for Excellence Grant. I would also like to thank my students in my Fall 2018 Linear Algebra class, Megan Bargstedt, Sarah Boggins, Samantha Chessen, Kasey Collins, Emily Cooper, Cecilia Dong, Matthew Foster, Michael Golaski, Eduardo Gonzalez, Hannah Noelle Griesbach, Joseph Keating, Yousaf Khan, Ryan Kugal, Carter Martin, McKenzie Miller, Amy Moore, David Norfleet, Timothy Redgrave, William Reynolds, Daniel Ryan, Isaac Sasser, Shannon Treacy, and Anne Williams, for helping me work through the manuscript before it went to publication.

1

Matrix Operations

Lab 0: An Introduction to MATLAB®

Introduction

MATLAB is a computer programming language that allows for matrix manipulation. MATLAB only recognizes certain commands that are relative to this program. Therefore you must type the commands as you see them. MATLAB is also case sensitive which means that if you see uppercase you must type uppercase and if you see lowercase you must type lowercase.

There are two ways that you can effectively use a MATLAB command. One way to run a MATLAB command is to type the command directly in the Command Window next to the $<<$ symbol and then hit return. This process is convenient when processing only one command at a time makes sense. When you wish to evaluate more than one command at once, it might make more sense to open a MATLAB script.

In order to start a MATLAB script, click on New Script in the tool bar and start typing your commands. In order to process your script after typing it, save your script and then click on the run button in the tool bar. It is also important to note that if you close MATLAB and come back to your work later, your work is not stored in the memory so it is a good idea to save your work so that you can reevaluate it later.

At any point if you are having difficulties, use the Help menu; it is very helpful.

For each lab, you will have to open a new MATLAB script file, also called an Editor file, and type all solutions in this document. So let's begin there.

Open a new MATLAB script file by choosing the New drop down menu followed by the script choice.

To save this file, choose Save in the drop down menu. Save this file as lab0.m.

Exercises:

a. Type: **x=6** and then press the Run button in the tool menu. Notice that the output will show up in the Command window (which is a separate window from the Editor).

b. On the next line in the Editor, type: **x=6;** and then press the run button. What is the difference between the output in part a and the output here? In each case, MATLAB stores 6 in the variable x.

c. Type: **x+5** and then press run.

d. Type: **disp('x+5')** and on the next line type **disp(x+5)**, then Run. Which x + 5 in the display statement actually produces the value 11?

In order to comment a line out in the editor put a % at the beginning of the line. That is, a line with a % at the beginning of it will not be processed by MATLAB when it is run. It might also be important to be able to clear the entire memory or a particular variable. If you wish to clear the entire memory, type **clear** or if you wish to clear a single variable, such as x, type **clear x**.

Basic Programming in MATLAB

In this section, we will assume a basic understanding of programming. We will discuss Tables of data, For Loops, and If-Then Statements here. Again, the Help menu is very helpful in this regard as well.

The colon, :, is one of the handiest symbols in MATLAB, as we will see. If we wish to create a Table of 11 points with values x, where x is the integers from -1 to 9, we would type **Name of Table= -1:9;**.

To identify the ith entry in table type **Name of Table(i)**. To identify the entries in the ith row in a table type **Name of Table(i,:)**.

The structure of a For Loop is:

$$for\ index = starting\ value : ending\ value$$
$$body\ statements;$$
$$end$$

All statements in the body of the For Loop must be separated by semicolons. We can create a table of values x^2 where x is the integers from -1 to 9, by starting with a table of zeros, typing **Name of Table=zeros(1,11)** and then creating the for loop,

$$for\ i = 1:11$$
$$Name(i) = (i-2)^2;$$
$$end$$

Exercises:

a. *Create a table named Table1 with entries equal to* $4*i$*, where i goes between 1 and 6.*

b. *Type Table1(5) in the Command Window to determine the* 5th *entry of this table.*

c. *Type and run the following code and determine what it does.*
$A = zeros(5);$
$for\ i = 1:5$
$A(i,i) = 1;$
end
$disp(A)$

The structure of the If-Then statement in MATLAB is:

$$if\ condition$$
$$body\ statements\ if\ condition\ is\ true;$$
$$end$$

Similarly, the structure of an If-Then-Else statement is :

$$if\ condition$$
$$body\ statements\ if\ condition\ is\ true;$$
$$else$$
$$body\ statements\ if\ condition\ is\ false;$$
$$end$$

Note there is also an elseif statement as well that may come in handy.

When stating conditions in your if-then statement you may have to test an equality. Here we have to distinguish in MATLAB between == and =. When you use the "=", single equals, this is an assignment where you are assigning a value. If you use the "==", double equals, MATLAB interprets this as a condition or test and returns True or False. A double equals should be used to test equality in an if-then condition.

Exercises:

a. *Type and run the following code and determine what it does.*

```
A = zeros(5);
for j = 1 : 5
for i = 1 : 5
A(i,i) = 1;
if i < j
A(i,j) = 2;
end;
end;
end;
disp(A)
```

In the above code, we call the pair of For Loops a *Nested For Loop* because one is inside the other.

b. *Write a nested for loop, with incremental variables i and j, which incorporates an if-then statement that creates a 5×5 table, A, whose entries are 1 when $i = 1$ or $j = 1$. All other entries of A should be zero.*

Lab 1: Matrix Basics and Operations

Introduction

A *matrix* is a rectangular array of numbers. The numbers in the array are called the *entries* of the matrix. $A = \begin{pmatrix} 1 & 2 & 3 \\ 4 & 5 & 6 \end{pmatrix}$ is a matrix.

The general form is
$$\begin{pmatrix} a_{11} & a_{12} & a_{13} & \cdots & a_{1n} \\ a_{21} & a_{22} & a_{23} & \cdots & a_{2n} \\ a_{31} & a_{32} & a_{33} & \cdots & a_{3n} \\ \vdots & \vdots & \ddots & \ddots & \vdots \\ a_{n1} & a_{n2} & a_{n3} & \cdots & a_{nn} \end{pmatrix}.$$

Defining a Matrix in MATLAB

Example: To define the matrix A above, type **A=[1 2 3; 4 5 6]**.

To find the dimensions of a matrix in MATLAB,

Type: **size(The Name of the Matrix)**

Exercises:

a. *Define the matrix* $B = \begin{pmatrix} 3 & 4 \\ 6 & 7 \\ 9 & 10 \end{pmatrix}.$

b. *Find the dimensions of the matrices A and B.*

c. *Explain what the dimensions of a matrix are telling you about the matrix.*

Operations on Matrices

Adding Two Matrices

To add two matrices together, type :
The Name of the Matrix1 + The Name of the Matrix2

Exercises:

a. *Find the sum* $A + B$. *You should get an error; explain why you think an error occurred.*

b. *Define matrix* $M = \begin{pmatrix} 4 & 5 & 1 \\ -1 & 3 & 2 \end{pmatrix}$. *Find* $A + M$ *and* $M + A$. *Is addition of matrices commutative?*

c. *Explain the process of matrix addition. What are the dimensions of the sum matrix. How would you take the difference of two matrices?*

Scalar Multiplication

To multiply a matrix by a constant c,

Type : **c*The Name of the Matrix**

Exercise: Multiply matrix A by the scalar 4. Is multiplication of a scalar from the left the same as multiplication of a scalar from the right? (i.e., does 4*A = A*4?)

Multiplying Two Matrices

To multiply two matrices together, type:

The Name of the Matrix1*The Name of the Matrix2

Exercises:

a. *Multiply matrix A on the right by matrix B.*

b. *Go to https://www.mathworks.com/matlabcentral/fileexchange/ 63993-matrix-multiplication-app and try some examples of matrix multiplication. Then describe the multiplication process.*

c. *Multiply matrix A on the left by matrix B. Was your description of the multiplication process correct? What are the dimensions of this matrix?*

d. *Multiply matrix A on the right by matrix M. You should get an error; explain why an error occurred.*

e. *Is matrix multiplication commutative? What has to be true about the dimensions of two matrices in order to multiply them together?*

The Transpose and Trace of a Matrix

The transpose of a matrix, A is denoted A^T. To take the transpose of a matrix,

Type : **The Name of the Matrix'**

Exercises:

a. *Take the transpose of matrix A and describe the transpose operation.*

b. *What are the dimensions of the matrix A^T?*

c. *What is $(A^T)^T$?*

d. *Calculate* $(A + M)^T$. *Does this equal* $A^T + M^T$?

e. *Calculate* $(AB)^T$. *Does this equal* $A^T B^T$?

f. *Calculate* $B^T A^T$. *What is this equal to?*

g. *Calculate* $(3A)^T$. *What is this equal to?*

h. *In the above exercises, you explored properties of the transpose of a matrix. Write down conjectures on the properties that you observed about the transpose.*

If the number of rows of a matrix is the same as the number of columns in that matrix we call the matrix a *square matrix*. The *trace* of a square matrix A, $tr(A)$, is a mapping taking a square matrix to a real number. To take the trace of a square matrix

Type: **trace(The Name of the Matrix)**

Define matrix $U = \begin{pmatrix} 1 & 2 & 3 \\ 4 & 5 & 0 \\ 0 & 2 & -1 \end{pmatrix}$ and $V = \begin{pmatrix} 1 & 0 & 0 \\ 4 & 3 & 0 \\ 0 & 0 & 2 \end{pmatrix}$.

Exercises:

a. *Calculate* $tr(U)$ *and* $tr(V)$ *and describe the trace operation.*

b. *Calculate* $tr(U + V)$. *Does this equal* $tr(U) + tr(V)$?

c. *Calculate* $tr(U^T)$. *Does this equal* $tr(U)$?

d. *Calculate* $tr(UV)$. *Does this equal* $tr(U)tr(V)$?

e. *Calculate* VU *and* $tr(VU)$. *Note that* $UV \neq VU$, *but does* $tr(UV) = tr(VU)$?

Lab 2: A Matrix Representation of Linear Systems

Introduction

You may remember back to the time when you were first learning algebra and your favorite math teacher challenged you to find a solution for x and y in a system with 2 equations with 2 unknown variables, such as $2x + 5y = 7$ and $4x + 2y = 10$. How did you do it?

My money is on solving for one variable in one equation, and substituting into the other. Or maybe you multiplied the first equation by a constant and subtracted the second from the first to solve, and then the story goes on. This method is fine and actually how we too will do it except in terms of matrices. The algorithm that we will use is called *Gaussian Elimination* (or *Gauss Jordan Elimination*).

Exercise: How many solutions are there to a system with 2 equations and 2 unknowns (in general)? How would you visualize these solutions?

A linear system in variables $x_1, x_2,...,x_k$ is of the form

$$
\begin{aligned}
a_{11}x_1 + a_{12}x_2 + ... + a_{1k}x_k &= b_1 \\
a_{21}x_1 + a_{22}x_2 + ... + a_{2k}x_k &= b_2 \\
a_{31}x_1 + a_{32}x_2 + ... + a_{3k}x_k &= b_3 \\
\vdots \quad \vdots \quad\quad \ddots \quad\quad \ddots \quad &\quad \vdots \\
a_{m1}x_1 + a_{m2}x_2 + ... + a_{mk}x_k &= b_m
\end{aligned}
$$

and can be written as the matrix equation

$$
\begin{pmatrix}
a_{11} & a_{12} & a_{13} & \cdots & a_{1k} \\
a_{21} & a_{22} & a_{23} & \cdots & a_{2k} \\
a_{31} & a_{32} & a_{33} & \cdots & a_{3k} \\
\vdots & \vdots & \ddots & \ddots & \vdots \\
a_{m1} & a_{m2} & a_{m3} & \cdots & a_{mk}
\end{pmatrix}
\begin{pmatrix}
x_1 \\ x_2 \\ x_3 \\ \vdots \\ x_k
\end{pmatrix}
=
\begin{pmatrix}
b_1 \\ b_2 \\ b_3 \\ \vdots \\ b_m
\end{pmatrix}.
$$

In the lab below, you will find all of the terms that you will need in order to move forward with Gaussian Elimination (or Gauss Jordan Elimination).

The Identity Matrix

The $n \times n$ *identity matrix* $I_n = \begin{pmatrix} 1 & 0 & 0 & \cdots & 0 \\ 0 & 1 & 0 & \cdots & 0 \\ 0 & 0 & 1 & \cdots & 0 \\ \vdots & \vdots & \ddots & \ddots & \vdots \\ 0 & 0 & 0 & \cdots & 1 \end{pmatrix}$. This matrix has

1's down the "main diagonal" and 0's everywhere else. The command for the $n \times n$ Identity Matrix is, **eye(n)**.

Row Echelon Form of a Matrix

A matrix is in *row echelon form* if
1) The first non-zero entry in each row is a one, called a *leading one*
2) Rows of all zeros are at the bottom of the matrix
3) All entries below leading ones are zeros
4) If $i < j$, the leading one in row i is to the left of the leading one in row j.

In addition, the matrix is in *reduced row echelon form* if
5) each column with a leading one has only zeros everywhere else.

Exercises:

a. *Use MATLAB to create a 4×4 Identity Matrix.*

b. *Given the system $2x + 5y = 7$ and $4x + 2y = 10$, create a coefficient matrix, A, using the coefficients of the variables.*

c. *Find the reduced row echelon form of A, type **rref(A)**.*

So how do we think about getting A into row echelon (Gaussian Elimination) or reduced row echelon form (Gauss Jordan Elimination)? We perform elementary row operations to the original matrix. And with every elementary row operation there is a corresponding elementary matrix.

Elementary Row Operations and the Corresponding Elementary Matrices

There are only three possible elementary row operations.

1. **Swap two rows in a matrix.** If you swap two rows in a 2×2 matrix, start with $I_2 = \begin{pmatrix} 1 & 0 \\ 0 & 1 \end{pmatrix}$ and perform this operation to get elementary matrix $E_1 = \begin{pmatrix} 0 & 1 \\ 1 & 0 \end{pmatrix}$.

2. **Multiply a row by a nonzero scalar (constant), k_1.** If you multiply

row two in a 2 × 2 matrix by $k_1 = -\frac{1}{8}$, start with $I_2 = \begin{pmatrix} 1 & 0 \\ 0 & 1 \end{pmatrix}$ and perform this operation to get elementary matrix $E_2 = \begin{pmatrix} 1 & 0 \\ 0 & -\frac{1}{8} \end{pmatrix}$.

3. **Add a nonzero multiple, k_2, of a row to another row.** If you add a multiple $k_2 = -2$ of row one to row two in a 2 × 2 matrix, start with $I_2 = \begin{pmatrix} 1 & 0 \\ 0 & 1 \end{pmatrix}$ and perform this operation to get elementary matrix $E_3 = \begin{pmatrix} 1 & 0 \\ -2 & 1 \end{pmatrix}$.

Exercises: Define $A = \begin{pmatrix} 2 & 5 \\ 4 & 2 \end{pmatrix}$.

a. *Calculate $E_1 A$, how is your new matrix related to A?*

b. *Calculate $E_2 A$, how is your new matrix related to A?*

c. *Calculate $E_3 A$, how is your new matrix related to A?*

d. *Calculate $E_5 E_4 E_2 E_3 A$, where $E_4 = \begin{pmatrix} 1 & -5 \\ 0 & 1 \end{pmatrix}$ and $E_5 = \begin{pmatrix} \frac{1}{2} & 0 \\ 0 & 1 \end{pmatrix}$, what is special about the matrix that you get?*

e. *Create a vector b with entries equal to the constants in the original system ($2x + 5y = 7$ and $4x + 2y = 10$), $b = \begin{pmatrix} 7 \\ 10 \end{pmatrix}$ and calculate $E_5 E_4 E_2 E_3 b$. If your original system is $Ax = b$ what is the new system after you perform the above operations? Use this to solve the original system of equations.*

f. *Choose another b and write down the system of equations, what is the solution to this system?*

g. *Let $M = \begin{pmatrix} 1 & 2 & 0 \\ 0 & 0 & 3 \\ 0 & 1 & 0 \end{pmatrix}$. Find elementary row operations and their corresponding elementary matrices such that when M is multiplied on the left by these matrices, the resulting matrix is I_3.*

h. *Solve the system $x + 2y = 4$, $3z = 6$, $y = 8$.*

i. *Let $M = \begin{pmatrix} 1 & 2 & 0 \\ 0 & 1 & 3 \\ 0 & 2 & 6 \end{pmatrix}$. What is the reduced row echelon form of M? Solve the system $x + 2y = 4$, $y + 3z = 6$, $2y + 6z = 18$.*

j. *Solve the system $x + 2y = 4, y + 3z = 6, 2y + 6z = 12$ and discuss how your result could be related to the reduced row echelon form of M.*

Lab 3: Powers, Inverses, and Special Matrices

Introduction

A *square matrix* is an $n \times n$ matrix.

If A is a square matrix and if a matrix B of the same size can be found such that $AB = BA = I$, then A is said to be *invertible* or *nonsingular* and B is called the *inverse* of A. If no such matrix B can be found, then A is said to be *singular*.

Powers of Matrices

Define the matrix $A = \begin{pmatrix} 1 & 2 & 0 \\ 2 & 1 & 0 \\ 0 & 1 & -2 \end{pmatrix}$, $B = \begin{pmatrix} 1 & -1 \\ 9 & 3 \\ 0 & 4 \end{pmatrix}$,

$M = \begin{pmatrix} 1 & 2 & 3 \\ 0 & 5 & 6 \\ 7 & 8 & 9 \end{pmatrix}$ and $P = \begin{pmatrix} 1 & 4 & 0 \\ 2 & 5 & 0 \\ 3 & 6 & 0 \end{pmatrix}$.

To determine the kth power of a matrix, where k is a positive integer value.

Type: **mpower(The Name of the Matrix,k)** or
The Name of the Matrix^k

Exercises:

a. *Calculate A^2. Is this the same as squaring all the entries in A? What is another way to express A^2?*

b. *Calculate B^2. An error occurred; determine why this error occurred. What property has to hold true in order to take the power of a matrix?*

c. *Determine what matrix A^0 is equal to.*

d. *Do the laws of exponents appear to hold for matrices? $A^r A^s = A^{(r+s)}$ and $(A^r)^s = A^{rs}$? Check these by example.*

Inverse of a Matrix

To determine the inverse of a matrix

Type: **inv(The Name of the Matrix)**

Exercises:

a. Find the inverse of A, A^{-1}. What are the dimensions of A^{-1}? What does AA^{-1} equal? What does $A^{-1}A$ equal?

b. Determine what matrix $(A^{-1})^{-1}$ is equal to.

c. Calculate $(AM)^{-1}$, $(MA)^{-1}$, $A^{-1}M^{-1}$, $M^{-1}A^{-1}$. Which of these matrices are equal?

d. Property : $(A^T)^{-1} = (A^{-1})^T$. Using the properties you have learned so far, which of the following are equal : $((AM)^T)^{-1}$, $((MA)^T)^{-1}$, $(A^{-1})^T(M^{-1})^T$, $(M^{-1})^T(A^{-1})^T$?

e. Find the inverse of P, P^{-1}. Can you explain why an error occurs? Note that the error is related to the matrix being singular.

Special Matrices

A square matrix, A, is *symmetric* if $A = A^T$.

A square matrix, A, is *diagonal*, if $A_{ij} = 0$ if $i \neq j$.

A square matrix, A, is *upper triangular* if $A_{ij} = 0$ when $i > j$ and is *lower triangular* if $A_{ij} = 0$ when $i < j$.

Exercises:

a. Determine what type of matrices $A + A^T$ and $M + M^T$ are and make a conjecture about a property related to your findings.

b. Define $Q = \begin{pmatrix} 1 & 0 \\ 2 & 3 \end{pmatrix}$ what type of matrix is Q^T? What type of matrix is Q^{-1}?

c. Find Q^2 and Q^3, what type of matrix is Q^k for any natural number k?

Theorems and Problems

For each of these statements, either prove that the statement is true or find a counter example that shows it is false.

Theorem 1. The inverse of an elementary matrix is an elementary matrix.
Theorem 2. If A is invertible then the reduced row echelon form of A is I.
Theorem 3. If the reduced row echelon form of A is I then A is invertible.
Theorem 4. A is a square invertible matrix if and only if A can be written as the product of elementary matrices.
Problem 5. If A is invertible then A^k is invertible for any natural number k.

Problem 6. If A is symmetric so is A^T.

Problem 7. If A is a symmetric invertible matrix then A^{-1} is symmetric.

Problem 8. If A and B are symmetric matrices of the same size then $A + B$ is symmetric.

Problem 9. If A and B are symmetric matrices of the same size then AB is symmetric.

Problem 10. If A is a square matrix then $A + A^T$ is symmetric.

Problem 11. The sum of upper triangular matrices is upper triangular.

Lab 4: Graph Theory and Adjacency Matrices

Basics of Graph Theory

A graph consists of vertices and edges. Each edge connects two vertices and we say that these two vertices are *adjacent*. An edge and a vertex on that edge are called *incident*. Given two vertices in a graph v_1 and v_2, the sequence of edges that are traversed in order to go from vertex v_1 to vertex v_2 is called a *path* between v_1 and v_2. Note that there is not necessarily a unique path between vertices in a graph.

A graph can be represented by an *adjacency matrix* where the ij^{th} entry of the adjacency matrix represents the adjacency between vertex i and vertex j. If vertex i and vertex j are adjacent then the ij^{th} entry is 1, otherwise it is 0.

It is also important to note that there are *directed graphs* and *undirected graphs*. A directed graph's edges are represented by arrows, and the edges of a directed graph can only be traversed in the direction that the arrow is pointing, similar to a one-way street. Here adjacency can also be recognized as being one directional. In an undirected graph, an edge is represented by a line segment and thus adjacency is symmetric.

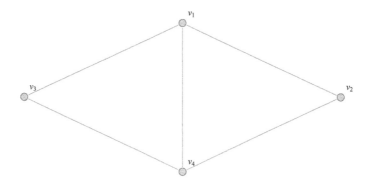

FIGURE 1.1

Exercises:

a. *Using the graph in Figure 1.1, create the adjacency matrix, A.*

b. *What type of special matrix is A?*

c. *To create a graph in MATLAB using your adjacency matrix, type :*

   plot(graph(The Name of the Matrix)).

Create the graph affiliated with adjacency matrix A using this command.

d. How many 1-step paths are there between vertex 1 and vertex 4? How many 2-step paths are there between vertex 1 and vertex 4?

e. Calculate A^2 and discuss how you can determine the number of 2-step paths between vertex 1 and vertex 4 using A^2.

f. The entries of the sum of what matrices would tell you how many paths of 3-steps or less go between vertex 1 and vertex 4?

An Application to Hospital Placements in Ghana

FIGURE 1.2: Map of Ghana

The country of Ghana has national hospitals located in three of its major cities, Accra, Cape Coast, and Techinan. However, many of its citizens from rural villages and small cities can never make it to these city hospitals based on road conditions and other infrastructure issues.

You are a member of the urban health and planning committee for Ghana and would like to strategically place one or two more hospitals in the cities of Dumbai, Damgo, Sunyani, or Kumasi so that all of the villages in the graphical representation of the map in Figure 1.3 can get to a national hospital without passing through more than one additional city. The black cities in Figure 1.3 are the cities in which a proposed hospital can be placed; the gray cities have no hospital and there is no proposal to place one there; and the white cities represent a city that currently has a national hospital.

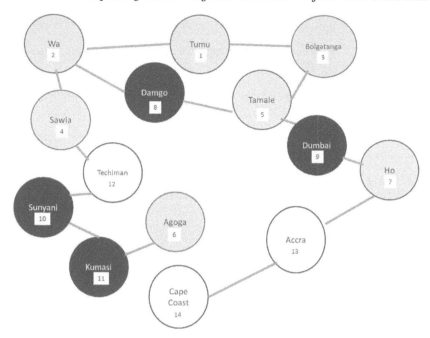

FIGURE 1.3: A graphical representation of the towns

Exercises:

a. *Is it currently possible to accomplish the goal of all of the villages on the map, represented by Figure 1.3, having access to a national hospital without passing through more than one additional city? If not what is the maximum number of cities that would have to be traversed in order for the entire population to get to a current hospital? Justify your answer using your knowledge of adjacency matrices and the graph in Figure 1.3.*

b. *What is the minimum number of additional hospitals that can be placed in proposed cities so that people in all of the villages and cities in the graph representation of the map, Figure 1.3, can go to an adjacent city or through at most one other city in order to reach a national hospital? Justify your answer with alterations to your adjacency matrix.*

Lab 5: Permutations and Determinants

Permutations

Given a set of elements, S, a *permutation* is an ordering of the elements of S. The demonstration `http://www.mathworks.com/matlabcentral/fileexchange/64083-permutations-app` shows the permutations as they relate to vertices on a graph. Use this demonstration to answer the following questions.

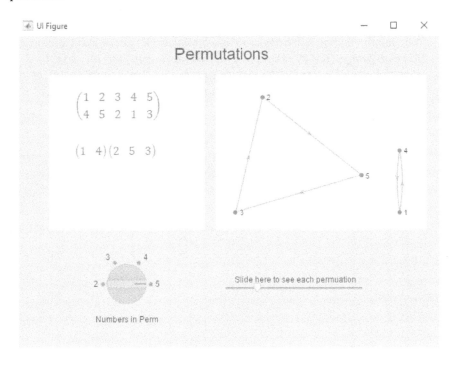

FIGURE 1.4

Example: Setting the number length (number of vertices) to 2. There are two notations used to represent the permutations: $\begin{pmatrix} 1 & 2 \\ 2 & 1 \end{pmatrix}$ and (12). Both of these representations say that the element in the 1st position goes to the 2nd position and the element in the 2nd position goes to position 1. Similarly, $\begin{pmatrix} 1 & 2 \\ 1 & 2 \end{pmatrix}$ and $(1)(2)$ leave the elements in the 1st and 2nd positions.

Exercises:

a. *Using the demonstration, write the permutations of 3 elements; how many are there?*

b. *How many permutations of 4 elements do you think there are?*

c. *The sign of a permutation is based on the number of switches that need to be made in order to get the numbers in order. For example, the sign of the permutation (132) is −1 since we need to make just one switch, of the numbers 2 and 3, to get back to (123). The sign of the permutation (312) is 1 since we can get to (123) in an even number of steps. Using the demonstration at* https://www.mathworks.com/matlabcentral/ fileexchange/64127-signed-determinant-app *set the size to 2 and step through the terms (the determinant of a 2 × 2 is the sum of these terms), discuss how the terms shown here relate to permutations of 2 elements.*

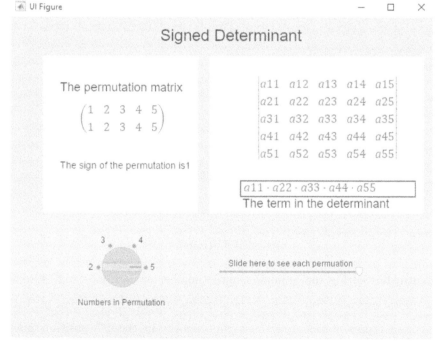

FIGURE 1.5

d. *What do you think the formula for a 3 × 3 determinant will look like? Use* your knowledge of permutations on 3 elements to argue your answer and then check your argument with the SignedDeterminant demonstration.

e. *Change the numbers in* $https://www.mathworks.com/matlabcentral/$ *fileexchange/64140-3x3determinant-app to see a trick for doing determinants of 3×3 matrices. Can you state a quick and easy way for doing 2×2 determinants?*

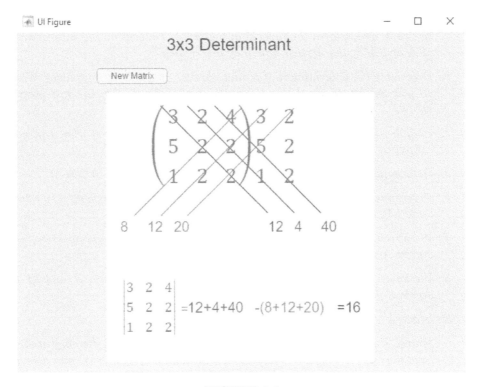

FIGURE 1.6

Determinants

The *determinant* of a matrix A is denoted $|A|$ or $det(A)$. To calculate

Type: **det(The Name of the Matrix)**

Exercises: Define $A = \begin{pmatrix} 2 & 2 & 0 \\ 2 & 1 & 0 \\ 0 & 1 & -2 \end{pmatrix}, B = \begin{pmatrix} 2 & 4 \\ 4 & 8 \end{pmatrix},$

$M = \begin{pmatrix} 1 & 2 & 3 \\ 0 & 5 & 6 \\ 7 & 8 & 9 \end{pmatrix}, P = \begin{pmatrix} 1 & 4 \\ 2 & 5 \end{pmatrix}, V = \begin{pmatrix} 1 & 0 & 0 \\ 4 & 3 & 0 \\ 0 & 0 & 2 \end{pmatrix},$ and

$W = \begin{pmatrix} 5 & 0 & 0 \\ 0 & 4 & 0 \\ 0 & 0 & -1 \end{pmatrix}.$

a. *In Lab 3, we explored the inverse of matrix A. Determine the determinant of A and A^{-1} and discuss how they are related.*

b. *Determine the determinant of B and whether or not B is invertible? What do you conjecture about the determinant of matrices that are not invertible?*

c. *Find $det(I_2)$ and $det(I_3)$. Based on these two calculations, what can you conjecture about the value of $det(I_n)$.*

d. *Determine $det(A^T)$ and discuss how this value is related to $det(A)$.*

e. *Determine $det(2A)$, $det(2P)$, $det(3A)$, $det(3P)$ and discuss how they relate to $det(A)$ and $det(P)$.*

f. *We already discovered that matrix multiplication is not commutative, use matrix A and M to decide if $det(AM) = det(MA)$.*

g. *We know that matrix addition is commutative; use matrix A and M to decide if $det(A + M) = det(M + A)$.*

h. *Is $det(A + M) = det(A) + det(M)$?*

i. *Matrix V is a lower triangular matrix and matrix W is a diagonal matrix (and thus also triangular); find the determinants of V and W and discuss how to find determinants of triangular matrices.*

j. *Calculate $\frac{(tr(P))^2 - tr(P^2)}{2}$ and $\frac{(tr(M))^3 - 3tr(M^2)tr(M) + 2tr(M^3)}{6}$ and determine how these quantities relate to $det(P)$ and $det(M)$, respectively.*

The quantities in part j are applications of the Cayley–Hamilton Theorem applied to 2×2 and 3×3 matrices.

Determinants of Elementary Matrices as They Relate to Invertible Matrices

Exercises: Define $E_1 = \begin{pmatrix} \frac{1}{2} & 0 \\ 0 & 1 \end{pmatrix}, E_2 = \begin{pmatrix} 1 & 0 \\ -4 & 1 \end{pmatrix},$ and

$E_3 = \begin{pmatrix} 0 & 1 & 0 \\ 1 & 0 & 0 \\ 0 & 0 & 1 \end{pmatrix}.$

a. If E_1 is an elementary matrix representing the operation of multiplying a row by a nonzero scalar, $k = \frac{1}{2}$, find $det(E_1)$. Make a conjecture about how this operation on a matrix effects the determinant of the matrix.

b. If E_2 is an elementary matrix representing the operation of adding a multiple of a row to another row, find $det(E_2)$. Make a conjecture about how this operation on a matrix effects the determinant of the matrix.

c. If E_3 is an elementary matrix representing the operation of switching two rows in a matrix, find $det(E_3)$. Make a conjecture about how this operation on a matrix effects the determinant of the matrix.

Theorems and Problems

For each of these statements, either prove that the statement is true or find a counter example that shows it is false.

Thereom 12. If $det(A)$ is not 0 then A is invertible.

Theorem 13. If A is invertible then $det(A)$ is not 0.

Problem 14. If A and B are invertible matrices of the same size then $A + B$ is invertible.

Theorem 15. If A is a square matrix then $det(A) = det(A^T)$.

Theorem 16. If A and B are matrices of the same size then A and B are invertible if and only if AB is invertible.

Lab 6: 4×4 Determinants and Beyond

In Lab 5, we discussed how to take the determinant of 2×2 and 3×3 matrices but what if you have larger matrices for which you have to take the determinant? One technique for finding determinants of larger matrices is called *Cofactor expansion.*

Let's use Cofactor expansion to find the determinant of $A = \begin{pmatrix} 1 & 1 & 0 & 0 \\ 1 & 2 & 1 & 0 \\ 2 & 1 & 3 & 1 \\ 0 & 0 & 1 & 4 \end{pmatrix}.$

To Do (Cofactor expansion) :

1. First choose a row or column of your matrix to expand upon. Any row or column will work but as you will see, choosing the row or column with the most 0's is the best choice.

2. Each entry in the matrix has a minor associated with it. The *minor* associated with entry i,j is the determinant of the matrix, M_{ij}, that is left when the ith row and j^{th} column are eliminated. So for example,
$$M_{11} = \det \begin{pmatrix} 2 & 1 & 0 \\ 1 & 3 & 1 \\ 0 & 1 & 4 \end{pmatrix}.$$

3. The determinant of an $n \times n$ matrix, with ij^{th} entry a_{ij}, when expanding about row i is $\sum_{j=1}^{n}(-1)^{(i+j)}a_{ij}M_{ij}$ and when expanding about column j is $\sum_{i=1}^{n}(-1)^{(i+j)}a_{ij}M_{ij}$.

Exercises:

a. *Calculate M_{41}, M_{42}, M_{43}, and M_{44} of A.*

b. *Use your minors M_{41} through M_{44} to find the determinant of A.*

c. *Expand about row 1 to find the determinant of A.*

d. *Define $B = \begin{pmatrix} 1 & 1 & 0 & 0 \\ 0 & 1 & 1 & 0 \\ 0 & -1 & 3 & 1 \\ 0 & 0 & 1 & 4 \end{pmatrix}$ and $P = \begin{pmatrix} 1 & 1 & 0 & 0 \\ 0 & 1 & 1 & 0 \\ 0 & 0 & 1 & 4 \\ 0 & 0 & 0 & -15 \end{pmatrix}$. Use co-factor expansion to find $|B|$ and your knowledge of determinants of upper triangular matrices from Lab 5 to find $|P|$.*

e. *Determine elementary matrices $E_1, E_2,$ and E_3 such that $E_3 E_2 E_1 B = P$.*

f. *In Lab 5, you conjectured about how row operations affect the determinant; use that knowledge along with properties of determinants, and part e., to find* $|B|$.

Project Set 1

Project 1: Lights Out

The 5×5 Lights Out game is a 5×5 grid of lights where all adjacent lights are connected. Buttons are adjacent if they are directly touching vertically or horizontally (not diagonally). In the Lights Out game, all buttons can be in one of two states, on or off. Pressing any button changes the state of that button and all adjacent buttons. The goal of this project is to create a matrix representation of the Lights Out game where all lights start on and need to be turned off. A picture of the Lights Out game with buttons labeled can be found in Table 1.1.

TABLE 1.1
5x5 Lights Out Grid

1	2	3	4	5
6	7	8	9	10
11	12	13	14	15
16	17	18	19	20
21	22	23	24	25

a. Note that since in the Lights Out game a button changes its own state when pressed, a button is adjacent to itself. Create the adjacency matrix, M, for the 5×5 game in Table 1.1.

b. A *row vector* is a $1 \times n$ matrix and a *column vector* is an $n \times 1$ matrix. If i is the initial state vector, what would the column vector i look like? Recall the goal is to determine if all lights can be turned off, starting with all lights on. (Use 0 for off and 1 for on.)

c. If f is the final state vector, determine f.

d. Write up your findings and supporting mathematical argument.

Project 2: Traveling Salesman Problem

Joe's Pizzeria wishes to send a single driver out from its main store who will make 4 deliveries and return to the store at the end of the route.

a. A *weighted adjacency matrix* is an adjacency matrix whose entries represent the weights of the edges between two adjacent vertices. For example, the weights in Figure 1.7 represent the time it takes to travel from one site, vertex, to another site. Create a weighted adjacency matrix, A, with the Joe's Pizzeria as vertex 1. A_{ij} should represent the time traveled by the driver between site i and site j.

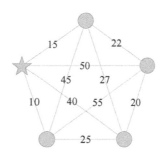

FIGURE 1.7: Map of delivery sites and Joe's Pizzeria denoted by a star

b. As mentioned before, the driver should start and end at the pizzeria while stopping at each of the delivery sites. The time of one such path is $A_{12} + A_{23} + A_{34} + A_{45} + A_{51}$. Calculate the time that the driver travels if the driver travels on this path. This path is using the off diagonal of A.

c. Other paths can easily be explored by looking at permutations of the rows of the matrix A. How many permutations are there?

d. The command **idx=perms([5 4 3 2 1])** will create a list of all permutations of the numbers 1 through 5, and the loop

$$\text{I=eye(5);}$$
$$\text{for c} = \text{1:120}$$
$$\text{P=I(idx(c,:),:)}$$
$$\text{disp(P*A*transpose(P))}$$
$$\text{end}$$

should produce all of the matrices which are permutations of the rows, and respective columns, of A.

If **P=I(idx(2,:),:)** and **B=P*A*transpose(P)**, use the off diagonal of B to determine another route that the driver can take and the time that the truck takes to traverse this route.

e. Write a small for loop utilizing the commands from part d. to find the path that gives the quickest route. Write up your findings and supporting mathematical argument.

Project 3: Paths in Nim

The demonstration found at https://www.mathworks.com/matlabcentral/
fileexchange/64175-counting-paths-of-nim-app shows the number of
paths (limited to a length of r) between point A in row 1 and B in row r
in the game of Nim with n rows. Your problem is to determine a matrix representation to determine the number of paths shown in this demonstration.

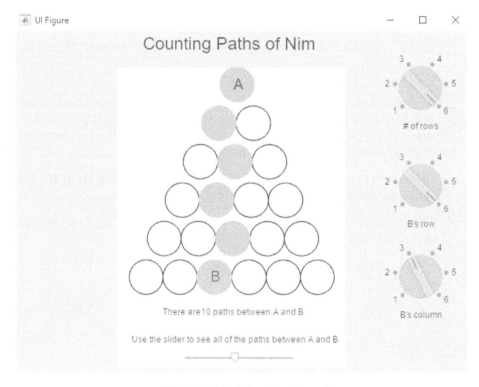

FIGURE 1.8: The Nim Board

a. If you did not care how long the path is from point A to point B (that is, the length is not limited by the number of rows, r), determine a matrix representation to count the number of 2-step paths, 3-step paths, and k-step paths. For simplicity allow n, the total number of rows in Nim, to be fixed at 5.

b. Make a conjecture about the number of k-step paths between a point A in row 1 and point B when B is position $(row, column) = (r, c)$ when there are 5 rows and in general n rows in the Nim game.

c. Using what you found, create a representation limiting the length of the path between A and B, as in the demonstration.

Project 4: Gaussian Elimination of a Square Matrix

Project 4 requires some programming in *MATLAB*. A small sample program is provided below which retrieves a matrix, A, and divides the first row by a_{11}.

```
prompt =' Input a matrix A';
A = input(prompt)
n = size(A);
temp = A(1,1);
for j = 1 : n(2)
A(1,j) = A(1,j)/temp;
end
disp(A)
```

a. Create a program (assuming that rows need not be swapped for Gaussian Elimination– that is assume no 0's will show up on the main diagonal) to get any square matrix A in row echelon form. Since we are only doing Gaussian Elimination of square matrices here, you may want to include an if-then statement that checks that the matrix is square.

b. Create a program where swaps are allowed to get any square matrix A in row echelon form.

c. Create a program where swaps are allowed to get any square matrix A in reduced row echelon form.

Project 5: Sports Ranking

In the 2013 season, the Big Ten football games in Table 1.2 occurred with W representing the winner. The question is how to rank these teams based on these games. The *dominance matrix*, A, is a matrix of zeros and ones where $A_{i,j} = 1$ if teams i and j played and team i won and $A_{i,j} = 0$ otherwise.

a. Create the dominance matrix and determine all one-step dominances for each team and one- and two-step dominances for each team combined.

b. Rank-order the teams by number of victories and by dominance.

c. Consider the dominance rankings of Minnesota and Michigan State. How is it possible that Minnesota has a higher dominance ranking than Michigan State while Minnesota has fewer victories than Michigan State?

TABLE 1.2

2013 Big Ten Results

Michigan State W – Indiana	Michigan State W – Purdue
Michigan State W – Illinois	Michigan State W – Iowa
Indiana W – Penn State	Penn State W – Michigan
Iowa W – Minnesota	Iowa W – Northwestern
Michigan W – Minnesota	Michigan W – Indiana
Minnesota W – Northwestern	Minnesota W – Wisconsin
Minnesota W – Nebraska	Nebraska W – Purdue
Nebraska W – Illinois	Ohio State W– Wisconsin
Ohio State W – Penn State	Ohio State W – Iowa
Ohio State W – Northwestern	Wisconsin W – Illinois
Wisconsin W – Northwestern	Wisconsin W – Purdue

d. Given that many times in a league every team does not necessarily play every other team, would ranking victories or dominance seem more reasonable for national rankings? How might one incorporate the score of the game into the dominance ranking as well?

Project 6: Archaeological Similarities, Applying Seriation

In archeology, *seriation* is a relative dating method in which assemblages or artifacts from numerous sites, in the same culture, are placed in chronological order. Most data that is collected is binary in nature where if an artifact, or record, possesses an identified trait, the artifact would be assigned a one for that trait and a zero otherwise.

In this project, there are 4 artifacts and 5 traits: Artifact A has Traits 1, 2, and 4. Artifact B has Traits 1, 3, 4, and 5, Artifact C has Traits 1, 2, 3, and 4, and Artifact D has Traits 1, 4, and 5.

a. Create a binary matrix, M, with rows representing artifacts and columns representing traits that the artifacts may possess.

b. $S = MM^T$ is called the *similarity matrix*. Find the similarity matrix and describe what $S_{i,i}$ and $S_{i,j}$ where $i \neq j$ represent.

c. $D = N - S$ where N is a matrix with all entries equal to n, where n is the number of traits. D is called the *dissimilarity matrix*. Many researchers who use seriation techniques attempt to find an ordering that minimizes some cost function. One cost function of interest is the number of dissimilarities. The dissimilarity between artifact i and artifact j is $D_{i,j}$, so the dissimilarity cost of an ordering of m artifacts $1,2,3,\ldots,m$ is $D(1,2,3,\ldots,m) = D_{1,2} + D_{2,3} + D_{3,4} + \cdots + D_{m-1,m} + D_{m,1}$. Find the dissimilarity matrix using matrix M and the dissimilarity cost for the ordering of artifacts $\{A,B,C,D\}$.

d. Find the dissimilarity cost for the ordering of artifacts $\{A,C,B,D\}$. How many unique orderings of these artifacts are there? Explore these different orderings and determine the ordering that minimizes the dissimilarity cost.

Project 7: Edge-Magic Graphs

A graph is called *edge-magic* if the edges can be labeled with positive integer weights such that (i) different edges have distinct weights, and (ii) the sum of the weights of edges incident to each vertex is the same; this sum is called the *magic constant.*

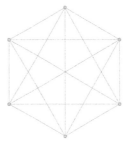

FIGURE 1.9

a. For the graph in Figure 1.9, create a system of linear equations that would determine the edge weights if the magic constant is 40.

b. Use your system from part a. to determine a solution, distinct edge weights of value 1 through 15, that produce a magic constant of 40. Recall that all edge weights must be nonzero.

c. The graph in Figure 1.9 is called the *complete graph* with 6 vertices, denoted K_6. In a complete graph with n vertices, denoted K_n, each pair of vertices is adjacent. Make a conjecture about edge-magic properties of K_n.

2

Invertibility

Lab 7: Singular or Nonsingular?
Why Singularity Matters

Introduction

Many topics that we discuss throughout this course will relate to the invertibility of a matrix. In Lab 3, we investigated some basic properties of matrices that were nonsingular, invertible, versus those that were singular, not invertible. In this lab, we will further look at inverses of matrices, how to calculate inverses, and how to use them to solve systems of equations.

So far in the labs, you may have noticed that there are many topics that are related. For example, how many different things can you think of at this point that are equivalent to saying that a square matrix, A, is invertible?

If A is an $n \times n$ matrix then the following are equivalent:

1. A is invertible.

2. $|A| \neq 0$.

3. The reduced row echelon form of A is I_n.

Finding Inverses

As we saw, in Lab 3, $MATLAB$ has built in functions for calculating inverses of matrices but let's look at how you would calculate an inverse.

If A is an $n \times n$ invertible matrix, to find the inverse, augment A with I_n, $(A|I_n)$, and perform elementary row operations, or left multiply A with elementary matrices, until the left-hand side is I_n. At this point, you have found the inverse of A on the right-hand side of your augmented matrix, $(I_n|A^{-1})$.

Exercises:

a. *Define* $B = (A|I_3)$ *where* $A = \begin{pmatrix} 2 & 2 & 0 \\ 2 & 1 & 0 \\ 0 & 1 & -2 \end{pmatrix}$.

b. Define $E_1 = \begin{pmatrix} \frac{1}{2} & 0 & 0 \\ 0 & 1 & 0 \\ 0 & 0 & 1 \end{pmatrix}$ and calculate $E_1 B$. What row operation does E_1 perform on B?

c. Continue to find elementary matrices, $E_2, \ldots, E_{k-1}, E_k$ (these are not unique) such that $E_k E_{k-1} \cdots E_2 E_1 B = (I|A^{-1})$. Another way to interpret this is that $E_k E_{k-1} \cdots E_2 E_1 A = I$.

d. Write A as a product of elementary matrices.

Using Inverses to Solve Systems of Linear Equations

In Lab 2, we used Gaussian Elimination (or Gauss Jordan Elimination) to solve the system $Ax = b$. We can also use our knowledge that $A^{-1}A = AA^{-1} = I$ to solve the system if A is invertible. Note $x = A^{-1}Ax = A^{-1}b$.

Thus, if A is invertible, the system $Ax = b$ has exactly one solution $x = A^{-1}b$.

Exercises:

a. Determine if $A = \begin{pmatrix} 1 & 2 & 0 \\ 0 & 1 & 3 \\ 1 & 2 & 6 \end{pmatrix}$ is invertible and use A^{-1} to solve the

system $Ax = b$ where $b = \begin{pmatrix} 7 \\ 10 \\ 0 \end{pmatrix}$.

b. Using matrix A and A^{-1} from part a., solve the system $Ax = \begin{pmatrix} 0 \\ 0 \\ 0 \end{pmatrix}$.

Part b. represents a special type of linear system. If the constants in the linear system are all 0 we call the linear system a *homogeneous linear system*. Homogeneous systems always have at least one solution. What is it?

The name of this solution is called the *trivial solution*.

Theorems and Problems

For each of these statements, either prove that the statement is true or find a counter example that shows it is false.

Problem 17. The inverse of a nonsingular upper triangular matrix is upper triangular.

Problem 18. The inverse of a nonsingular diagonal matrix is diagonal.

Problem 19. $|A^{-1}| = \frac{1}{|A|}$.

Theorem 20. A is invertible if and only if A can be written as a product of elementary matrices.

Theorem 21. If A is an $n \times n$ invertible matrix then the system $Ax = b$ has exactly one solution for all $n \times 1$ vectors b.

Theorem 22. If A is an $n \times n$ matrix and the system $Ax = b$ is *consistent* (has at least one solution) for all $n \times 1$ vectors b then A is invertible.

Problem 22. If $ad - bc \neq 0$ then $\begin{pmatrix} a & b \\ c & d \end{pmatrix}^{-1} = \frac{1}{(ad-bc)} \begin{pmatrix} d & -b \\ -c & a \end{pmatrix}$.

Theorem 24. A is an $n \times n$ invertible matrix if and only if the system $Ax = 0$ has only the trivial solution.

What can you add now to your list of statements that is equivalent to the statement A is invertible?

If A is an $n \times n$ matrix the following are equivalent statements:

1. A is invertible.

2. $|A| \neq 0$.

3. The reduced row echelon form of A is I_n.

4. A can be written as a product of elementary matrices.

5. The system $Ax = b$ has exactly one solution for all $n \times 1$ vectors b.

6. The system $Ax = 0$ has only the trivial solution.

Lab 8: Mod It Out, Matrices with Entries in Z_p

Integers Modulo p

All of the matrices that we have been dealing with thus far have entries that are real numbers with addition and scalar multiplication defined as traditional addition and scalar multiplication of the reals. But what if we only work with integers and redefine addition and scalar multiplication?

If x and y are integers, we say that x and y are *congruent modulo p*, written $x \equiv y(mod\ p)$, if $x - y$ is an integer multiple of p, where p is an integer. For example,

0 and 6 are congruent modulo 3, $0 \equiv 6(mod\ 3)$, since $0 - 6$ is an integer multiple of 3,
12 and 7 are congruent modulo 5, $12 \equiv 7(mod\ 5)$, since $12 - 7$ is an integer multiple of 5,
-1 and 6 are congruent modulo 7, $-1 \equiv 6(mod\ 7)$, since $-1 - 6$ is an integer multiple of 7.

Exercises:

a. *To calculate $y(mod\ p)$, type **mod(y,p)**. Use this command to find $1(mod\ 3), 2(mod\ 3), 3(mod\ 3),$ and $4(mod\ 3)$.*

b. *What integer answers can occur when you do Mod 3? What about Mod 5?*

c. *The set of integers that can occur Mod 3 is called the Integers Mod 3 or Z_3 and similarly those that can occur Mod 5 are called Integers Mod 5 (or Z_5). Write the elements of Z_3 and Z_5 in set notation.*

Additive and Multiplicative Inverses in Z_p

When dealing with real numbers we think of the additive identity as 0 since for all real numbers x, $x + 0 = x$. We also denote the additive inverse as $-x$ where $x + (-x) = 0$. If 0 is the additive identity in modular arithmetic let's find additive inverses.

Example: In $Z_3, 1 + 2 \equiv 0(mod\ 3)$ so 1 and 2 are additive inverses in Z_3.

In the reals, the multiplicative inverse is 1 since for all real number x, $x \cdot 1 = 1 \cdot x = x$. We also denote the multiplicative inverse of x as $\frac{1}{x}$ in the reals since $x \cdot \frac{1}{x} = 1$. 1 is also the multiplicative identity in modular arithmetic, so how do we think of multiplicative inverses in Z_p?

Example: In Z_3, $1 \cdot 1 \equiv 1(mod\ 3)$ and $2 \cdot 2 \equiv 1(mod\ 3)$ so 1 is its own multiplicative inverse and 2 is its own multiplicative inverse in Z_3.

Note that if p is not prime, the elements of Z_p may not have a multiplicative inverse. For example, in Z_6, $2 \cdot 0 \equiv 0(mod\ 6)$, $2 \cdot 1 \equiv 2(mod\ 6)$, $2 \cdot 2 \equiv 4(mod\ 6)$, $2 \cdot 3 \equiv 0(mod\ 6)$, $2 \cdot 4 \equiv 2(mod\ 6)$, $2 \cdot 5 \equiv 4(mod\ 6)$ and thus 2 does not have a multiplicative inverse in Z_6.

Exercises:

 a. *Find the additive inverses of all elements of Z_5.*

 b. *Find the multiplicative inverses of all of the nonzero elements of Z_5.*

Matrices with Entries in Z_p

When adding two matrices, multiplying two matrices, multiplying a matrix by a scalar, or finding the determinant of a matrix, do these calculations as if the entries are in the reals and then convert the values to integers modulo p, where p is an integer.

Examples:

$$\begin{pmatrix} 1 & 2 \\ 3 & 4 \end{pmatrix} + \begin{pmatrix} 5 & 6 \\ 7 & 8 \end{pmatrix} = \begin{pmatrix} 6 & 8 \\ 10 & 12 \end{pmatrix} \equiv \begin{pmatrix} 0 & 2 \\ 1 & 0 \end{pmatrix} (mod\ 3).$$

$$\begin{pmatrix} 1 & 2 \\ 3 & 4 \end{pmatrix} \cdot \begin{pmatrix} 5 & 6 \\ 7 & 8 \end{pmatrix} = \begin{pmatrix} 19 & 22 \\ 43 & 50 \end{pmatrix} \equiv \begin{pmatrix} 1 & 1 \\ 1 & 2 \end{pmatrix} (mod\ 3).$$

$$2 \begin{pmatrix} 1 & 2 \\ 3 & 4 \end{pmatrix} = \begin{pmatrix} 2 & 4 \\ 6 & 8 \end{pmatrix} \equiv \begin{pmatrix} 2 & 1 \\ 0 & 2 \end{pmatrix} (mod\ 3).$$

$$det \begin{pmatrix} 1 & 2 \\ 3 & 4 \end{pmatrix} = -2 \equiv 1(mod\ 3).$$

Elementary row operations are the same on matrices with entries in integers modulo p as they are with entries which are real. Keep the following points in mind:

1. Use additive inverses modulo p when adding a multiple of a row to another row to get zeros everywhere except where leading ones are located.

2. Use multiplicative inverses modulo p when trying to make a number the leading one in a row.

3. When multiplying a row by a scalar, the scalars that you should use are those that are in Z_p.

Example: Find $A^{-1} = \begin{pmatrix} 1 & 2 \\ 1 & 1 \end{pmatrix}^{-1} (mod\ 3)$. Since $det \begin{pmatrix} 1 & 2 \\ 1 & 1 \end{pmatrix} \not\equiv 0(mod\ 3)$ we know the matrix is invertible modulo 3.

1. Augment the matrix $(A|I_2) = \begin{pmatrix} 1 & 2 & 1 & 0 \\ 1 & 1 & 0 & 1 \end{pmatrix}$.

2. Take $2 \times \text{Row } 1 + \text{Row } 2 \Longrightarrow \begin{pmatrix} 1 & 2 & 1 & 0 \\ 3 & 5 & 2 & 1 \end{pmatrix} \equiv \begin{pmatrix} 1 & 2 & 1 & 0 \\ 0 & 2 & 2 & 1 \end{pmatrix} (mod\ 3)$.

3. $2 \times \text{Row } 2 \Longrightarrow \begin{pmatrix} 1 & 2 & 1 & 0 \\ 0 & 4 & 4 & 2 \end{pmatrix} \equiv \begin{pmatrix} 1 & 2 & 1 & 0 \\ 0 & 1 & 1 & 2 \end{pmatrix} (mod\ 3)$.

4. $\text{Row } 2 + \text{Row } 1 \Longrightarrow \begin{pmatrix} 1 & 3 & 2 & 2 \\ 0 & 1 & 1 & 2 \end{pmatrix} \equiv \begin{pmatrix} 1 & 0 & 2 & 2 \\ 0 & 1 & 1 & 2 \end{pmatrix} (mod\ 3)$ and

$A^{-1} = \begin{pmatrix} 2 & 2 \\ 1 & 2 \end{pmatrix}$ in Z_3.

Exercises:
Use the demonstration $https://www.mathworks.com/matlabcentral/$
$fileexchange/65139\text{-}inverse\text{-}and\text{-}nullspaces\text{-}in\text{-}gf\text{-}p$ to help find the
inverse of a matrix modulo p.

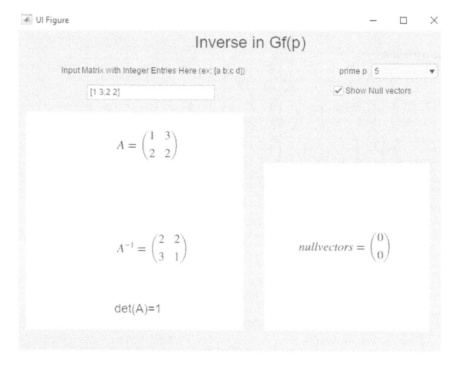

FIGURE 2.1

a. *Is* $\begin{pmatrix} 1 & 2 \\ 1 & 1 \end{pmatrix}$ *invertible modulo 2? Is it invertible modulo 5?*

b. *Solve the system $Ax = b$, where $A = \begin{pmatrix} 1 & 2 \\ 1 & 1 \end{pmatrix}$ and $b = \begin{pmatrix} 4 \\ 1 \end{pmatrix}$ modulo 2. Solve the same system modulo 5.*

Lab 9: It's a Complex World

Introduction

In Lab 8, we explored invertibility using modular arithmetic. Here we will be exploring the idea of matrices with entries which are complex numbers.

A complex number $z = a + bi$ has a real part a and complex part b multiplied by i where $\sqrt{-1} = i$. Use **i** in *MATLAB*. Every complex number has a complex conjugate. The *complex conjugate* of $z = a + bi$ is $\overline{z} = a - bi$.

Note that $z \cdot \overline{z} = a^2 + b^2$ is called the *magnitude* of z.

If a matrix has complex entries, $A = \begin{pmatrix} 2 + 3i & 7 - 8i \\ 5 - i & 2 \end{pmatrix}$, the complex conjugate of A is $\overline{A} = \begin{pmatrix} 2 - 3i & 7 + 8i \\ 5 + i & 2 \end{pmatrix}$.

Define A from above and $B = \begin{pmatrix} i & 1 \\ 0 & -i \end{pmatrix}$.

Exercises: To calculate the conjugate transpose, $\overline{A}^T$, which is the complex conjugate and transpose of a matrix,

Type: **ctranspose(The Name of the Matrix)**

Use this command for the following exercises.

a. *Calculate $\overline{A}^T$.*

b. *What matrix is $\overline{\overline{A}}$ equal to?*

c. *Does $\overline{A + B} = \overline{A} + \overline{B}$?*

d. *Does $\overline{AB} = \overline{A}\ \overline{B}$?*

Eigenvalues

For each $n \times n$ matrix A, we can calculate the *eigenvalues* of A by finding the values for λ such that $Ax = \lambda x$. The $\lambda's$ are the eigenvalues for A and each eigenvalue has a corresponding *eigenvector* x.

Another way to find the eigenvalues is to solve for λ in the *characteristic equation* $|A - \lambda I| = 0$.

Exercises: To calculate the eigenvalues of a matrix,

Type: **eig(The Name of the Matrix)**

a. Find the eigenvalues of $\begin{pmatrix} 2 & 0 \\ 0 & 3 \end{pmatrix}$. Use your result to make a conjecture about the values of eigenvalues of any diagonal matrix in general.

b. Make a conjecture about the values of eigenvalues of singular matrices. If you are unsure, try some examples.

c. Calculate the eigenvalues of $A = \begin{pmatrix} 1 & 4 \\ 2 & 3 \end{pmatrix}$ and the eigenvalues of A^T. What is the relationship between the eigenvalues of these two matrices?

d. Calculate the eigenvalues of $A = \begin{pmatrix} 1 & -i \\ 2i & i \end{pmatrix}$ and the eigenvalues of $\overline{A}^T$. What is the relationship between the eigenvalues of these two matrices? Explain why you saw similar properties in part c.

e. A square matrix A is called Hermitian if $\overline{A}^T = A$. Give an example of a Hermitian matrix with complex entries.

f. Using your example from part e, what can you conjecture about the eigenvalues of Hermitian matrices?

g. A square matrix A is called Unitary if $\overline{A}^T A = A\overline{A}^T = I$. Give an example of a unitary matrix with complex entries.

h. Using your example for part g, what can you conjecture about the eigenvalues of unitary matrices? If you are uncertain, try a few more examples.

Theorems and Problems

For each of these statements, either prove that the statement is true or find a counter example that shows it is false.

Problem 25. $|\overline{A}| = \overline{|A|}$.

Problem 26. If A is invertible, then $(\overline{A})^{-1} = \overline{A^{-1}}$.

Problem 27. If c is a complex number, then $\overline{cA} = c\overline{A}$.

Problem 28. The eigenvalues of a diagonal matrix are the entries on the main diagonal.

Theorem 29. All eigenvalues of Hermitian matrices are real numbers.

Theorem 30. The complex conjugate of a Hermitian matrix is a Hermitian matrix.

Theorem 31. A is Unitary if and only if $A^{-1} = \overline{A}^T$.

Problem 32. λ is an eigenvalue of matrix A if and only if it is an eigenvalue of matrix $\overline{A}^T$.

Lab 10: Declaring Independence: Is It Linear?

Linear Combinations

If $S = \{v_1, v_2, \ldots, v_m\}$ is a set of vectors and there exists scalars $k_1, k_2, \ldots, k_m$ such that vector $w = k_1 v_1 + k_2 v_2 + \cdots + k_m v_m$, we say that w can be written as a *linear combination* of $v_1, v_2, \ldots, v_m$.

If v_i is in R^m, we can think of the above definition of linear combination as

$$\begin{pmatrix} w_1 \\ w_2 \\ \vdots \\ w_m \end{pmatrix} = \begin{pmatrix} v_{11} & v_{21} & v_{31} & \cdots & v_{m1} \\ v_{12} & v_{22} & v_{32} & \cdots & v_{m2} \\ \vdots & \vdots & \vdots & \ddots & \vdots \\ v_{1m} & v_{2m} & v_{3m} & \cdots & v_{mm} \end{pmatrix} \begin{pmatrix} k_1 \\ k_2 \\ \vdots \\ k_m \end{pmatrix} \quad \text{where } w_i \text{ is the } i^{th}$$

entry of vector w and v_{ij} is the jth entry of v_i.

So determining if w can be written as a linear combination of $v_1, v_2, \ldots, v_m$ is equivalent to determining if there is a solution to the system $Ax = w$, where x is the vector of values k_i and the ith column of A is the vector v_i. If no solution exists then w cannot be written as a linear combination of $v_1, v_2, \ldots, v_m$.

Exercises:

a. *Can the vector* $w = (1,2,3)$ *be written as a linear combination of* $v_1 = (1,0,0), v_2 = (0,1,0),$ *and* $v_3 = (0,0,1)$?

b. *Can the vector* $w = (1,2,3)$ *be written as a linear combination of* $v_1 = (1,1,2), v_2 = (5,6,0),$ *and* $v_3 = (9,10,-2)$?

Linear Independence

If $S = \{v_1, v_2, \ldots, v_m\}$ is a set of vectors, we say that S is *linearly independent* if the homogeneous system $0 = k_1 v_1 + k_2 v_2 + \cdots + k_m v_m$ has only the trivial solution. Otherwise we say that the set is *linearly dependent*.

Exercises:

a. *Give an example of a set of vectors in* R^2 *that is linear dependent.*

b. *To visualize the set of vectors* $\{\{1,1,1\},\{1,2,3\},\{2,10,5\}\}$ *in* R^3 *type the following set of commands. Once you enter these commands, you can grab the diagram to rotate your view using the Rotate 3D in the Tools drop down menu. Type:*

$$t = 0:.1:4;$$
$$X = [0\ 2\ 2\ 0];$$
$$Y = [0\ 2\ 4\ 0];$$
$$Z = [0\ 2\ 6\ 0];$$

figure

fill3(X,Y,Z,'yellow')

hold on

*plot3(2 * t,10 * t,5 * t,'green',t,2 * t,3 * t,'red',t,t,t,'blue')*

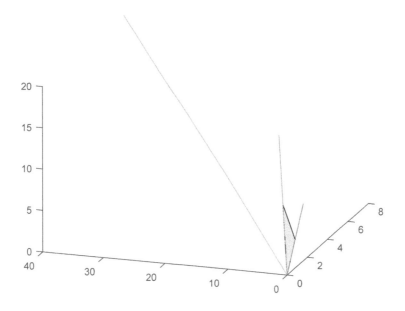

FIGURE 2.2

Is the set of vectors in Figure 2.2 linearly independent or linearly dependent? How would we visualize three linearly dependent vectors in R^3?

c. *Is the set of functions $\{1, \sin^2(x), \cos^2(x)\}$ linearly independent? Explain your answer.*

d. *How can we use techniques similar to those used to determine if a set in*

R^n is linearly independent to determine if a set in $M_{2,2}$, the set of 2×2 matrices, is linearly independent?

e. *Determine if the set* $\{ \begin{pmatrix} 1 & -1 \\ 2 & 0 \end{pmatrix}, \begin{pmatrix} 1 & 0 \\ 0 & 2 \end{pmatrix}, \begin{pmatrix} 2 & 0 \\ 0 & 1 \end{pmatrix}, \begin{pmatrix} 2 & -1 \\ 0 & 1 \end{pmatrix} \}$ *is linearly independent.*

Span

A set of vectors, S, is said to *span* V if every vector in V can be written as a linear combination of vectors in S. For example, every vector in R^2 looks like an ordered pair (a,b) and can be written as a linear combination of the vectors in the set $\{(1,0),(0,1)\}$. $(a,b) = a(1,0) + b(0,1)$.

Exercises:

a. *Graphically represent the span of the vector* $(1,2)$.

b. *Graphically represent the span of the set of vectors* $\{(1,2),(4,5)\}$.

c. *If you create a matrix with the vectors of a set as the rows (or columns) of your matrix, how would you determine if the vectors span* R^2?

d. *Does* $\{(1,2),(4,5)\}$ *span* R^2? *If so we say* $span(\{(1,2),(4,5)\}) = R^2$.

e. *Does* $span(\{(1,2),(4,5)\}) = span(\{(2,3),(4,5)\})$? *How would you determine if these spans are equal?*

f. *Does* $\{ \begin{pmatrix} 1 & -1 \\ 2 & 0 \end{pmatrix}, \begin{pmatrix} 1 & 0 \\ 0 & 2 \end{pmatrix}, \begin{pmatrix} 2 & 0 \\ 0 & 1 \end{pmatrix}, \begin{pmatrix} 2 & -1 \\ 0 & 1 \end{pmatrix} \}$ *span* $M_{2,2}$?

Theorems and Problems

For each of these statements, either prove that the statement is true or find a counter example that shows it is false.

Theorem 33. If A is invertible then the rows of A are linearly independent.
Theorem 34. If A is invertible then the columns of A are linearly independent.
Problem 35. A set of vectors with only two vectors in it is linearly dependent if one is a scalar multiple of the other.
Problem 36. A set of vectors is linear dependent if it contains the zero vector.
Theorem 37. If A is an $n \times n$ invertible matrix then the rows of A span R^n.
Theorem 38. If A is an $n \times n$ invertible matrix then the columns of A span R^n.

Project Set 2

Project 1: Lights Out

The 5×5 Lights Out game was introduced in Project Set 1 where you created the adjacency matrix, initial state vector, and final state vector. Recall that the goal of this game is, if all lights start on, to turn all lights off. We will move toward finding a solution to this problem.

TABLE 2.1
5x5 Lights Out Grid

1	2	3	4	5
6	7	8	9	10
11	12	13	14	15
16	17	18	19	20
21	22	23	24	25

a. It does not matter in what order buttons are pushed, so imagine that all of the buttons you are going to push will be pushed all at once. A push vector p can be created, where 0 represents a button which is not pushed and 1 represents a button which is pushed. Create a push vector where buttons 1, 8, and 25 are pushed and all others are not.

b. With M as the adjacency matrix, what does Mp represent in general? Calculate Mp using the vector p from part a. Is this push vector a solution to the 5×5 Lights Out game?

c. The goal is to find a solution (i.e., a push vector) such that $Mp + i = f$ with the initial state vector, i, and final state vector, f, that you defined in Project Set 1. Do you think that the game has a solution and if so, what is it?

d. Now assume that the buttons in the 5×5 Lights Out game can take on three states, 0, 1, and 2, and the goal of the game is to go from an initial state of all lights in state 0 and end with all lights in state 1. How will your process in finding a solution change with this new version of the game? Does this game have a solution and if so what is it?

e. Write up your findings and supporting mathematical argument.

Project 2: Hill Ciphers

A cipher is a coding system. In this project we introduce a basic cipher called *Hill Ciphering*. In Hill Ciphering, each letter is represented by a number. We will also add into the coding some punctuation; see below.

A	B	C	D	E	F	G	H	I	J	K	L	M	N	O	P	Q
0	1	2	3	4	5	6	7	8	9	10	11	12	13	14	15	16

R	S	T	U	V	W	X	Y	Z		.	?
17	18	19	20	21	22	23	24	25	26	27	28

Enciphering : To encipher a message, choose a 2×2 matrix which is invertible modulo 29, called your *encryption matrix*. You then must separate your message into 2×1 vectors which will then be multiplied by your encryption matrix modulo 29. If your code has an odd number of letters then repeat the last letter.

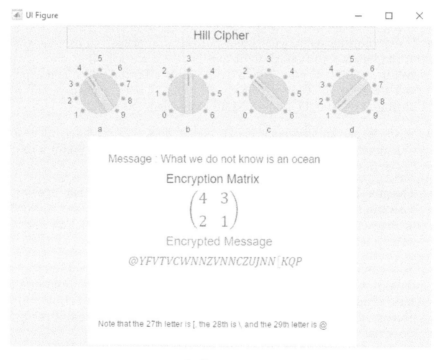

FIGURE 2.3

Deciphering: In order to decipher a code, you need to be given a 2×2 *decryption matrix*. This is the matrix that the code was enciphered with modulo 29. Thus in order to decipher the code you will have to separate the message into 2×1 vectors which will then be multiplied by the inverse of the decryption matrix modulo 29. To see examples of encryption and decryption see the demonstration `https://www.mathworks.com/matlabcentral/fileexchange/63769-hill-cipher-app`.

a. Create a 2×2 matrix which is invertible modulo 29 and a message that you would like to encipher.

b. Encipher your message from part a.

c. Use the decryption matrix $\begin{pmatrix} 1 & 4 \\ 13 & 6 \end{pmatrix}$ to decipher the code

$$\text{MLS?AJGN.LHP}$$

d. Write up your findings and supporting mathematical argument.

Project 3: Leontief Closed Production Model

In a Closed Economy Leontief Model, each industry has a production level p_i, and each industry i has a consumption level for product j, c_{ij}. If the economy is balanced, the total production of each industry must be equal to its total consumption and thus producing a linear system of equations.

$$
\begin{aligned}
c_{11}p_1 + c_{12}p_2 + \ldots + c_{1k}p_k &= p_1 \\
c_{21}p_1 + c_{22}p_2 + \ldots + c_{2k}p_k &= p_2 \\
c_{31}p_1 + c_{32}p_2 + \ldots + c_{3k}p_k &= p_3 \\
\vdots \quad \vdots \quad \ddots \quad \ddots \quad \vdots \\
c_{m1}p_1 + c_{m2}p_2 + \ldots + c_{mk}p_k &= p_m.
\end{aligned}
$$

Thus the goal is to find the amount of production that maintains the economy, solving for p in $Cp = p$. One can think of this also as finding the eigenvector associated with eigenvalue 1; more explanation on this to come in later labs. The matrix C is called the consumption or input-output matrix.

The Problem: According to the United Nations International Merchandise Trade Statistics, each of the following countries, China, India, and Singapore, provides large amounts of exports to the others. Table 2.2 shows the units of trade between countries that we will use for this problem.

Assuming a closed economy between these three countries, what ratio of commodities should each country produce in order to keep the economy stable?

TABLE 2.2

Trade between China, India, and Singapore

Consumption Country	China	India	Singapore
China	0.46	0.2	0.58
India	0.36	0.7	0.13
Singapore	0.18	0.1	0.29

Project 4: Modeling Influenza

At Malady College, a college campus of 5000 students, the spread of influenza is rampant. In this problem, we will call each student either susceptible or infected with influenza, and if the student is not infected they are susceptible. During any given year, the percentage of the U.S. population that will get the flu, on average, each year is between 5% and 20%. At Malady, if a student is not infected with influenza the chance that they will catch the flu on any given day is 16%, and if a student has the flu the chance that they will recover and return to susceptible on any given day is 40%.

a. Create a matrix, A, (called the *transition matrix*) whose columns represent the current state a student, either susceptible or infected, may be in and whose rows represent the state of a student, either susceptible or infected, tomorrow, and where A_{ij} is the probability of a student going from current state j today to state i tomorrow.

b. If 100 students have the flu initially, how many students have the flu on the second day? How many students have the flu on the tenth day?

c. How many days does it take for the number of students with the flu to stabilize?

d. How many students have the flu initially if there are 1400 students with the flu on the third day?

Project 5: Diagonalization of a Square Matrix

An $n \times n$ matrix A is *diagonalizable* if there exists matrix P such that $D = P^{-1}AP$, and D is a diagonal matrix. In addition, matrix A is diagonalizable if it has n linearly independent eigenvectors. The n linear independent eigenvectors are the n columns of P. Create a program that determines if any square matrix A is diagonalizable and diagonalizes A if it is diagonalizable.

Project 6: Balancing Chemical Equations

In a chemical equation for a reaction, the substances reacting (the reactants) are on the left side of the equation with an arrow pointing to the substances being formed on the right side of the equation (the products). The law of conservation of mass states that no atoms can be created or destroyed in a chemical reaction, so the number of atoms that are present in the reactants in a chemical reaction has to balance the number of atoms that are present in the products of that reaction. Thus, in order to write a correct chemical equation, we must balance all of the atoms on the left side of the reaction with the atoms on the right side. Each of the reactants and products has a vector affiliated with it, where the number of atoms of each element present are the entries of the individual vectors. For example in the unbalanced equation $x_1 N_2 + x_2 H_2 \longrightarrow x_3 N H_3$, we wish to find x_1, x_2, and x_3 that balance the equation. We define each reactant\product in terms of a vector representing the numbers of atoms of each element in the compound.

$$x_1 \begin{pmatrix} 2 \\ 0 \end{pmatrix} + x_2 \begin{pmatrix} 0 \\ 2 \end{pmatrix} = x_3 \begin{pmatrix} 1 \\ 3 \end{pmatrix}.$$

There are two ways to solve this system:

1. This is equivalent to solving the system $Cx = 0$ where C is the coefficient matrix $\begin{pmatrix} 2 & 0 & -1 \\ 0 & 2 & -3 \end{pmatrix}$. Here since C is not square it is not invertible but we can choose $x_3 = t$ and then $x_2 = \frac{3}{2}t$ and $x_1 = \frac{1}{2}t$. Letting $x_3 = 4$ we get a balanced equation $2N_2 + 6H_2 \longrightarrow 4NH_3$.

2. Define coefficient matrix $C = \begin{pmatrix} 2 & 0 \\ 0 & 2 \end{pmatrix}$ and $b = \begin{pmatrix} 1 \\ 3 \end{pmatrix}$ then $\begin{pmatrix} x_1 \\ x_2 \end{pmatrix} = det(C)C^{-1}b = \begin{pmatrix} 2 \\ 6 \end{pmatrix}$ and $x_3 = det(C) = 4$.

Now it is your turn to try it.

a. Balance equation $FeCl_2 + Na_3(PO_4) \longrightarrow Fe_3(PO_4)_2 + NaCl$ by solving the system $Cx = 0$ where C is the coefficient matrix.

b. Balance the $x_1 Cu_2 S + x_2 O_2 \longrightarrow x_3 Cu + x_4 SO_2$ by creating a 3×3 coefficient matrix C and solve the system $Cx = b$ with $\begin{pmatrix} x_1 \\ x_2 \\ x_3 \end{pmatrix} = det(C)C^{-1}b$ and $x_4 = det(C)$.

Project 7: Magic Squares

A *magic square* is an arrangement of positive integers in a square grid, where the numbers in each row, in each column, and the numbers in the main diagonals, all add up to the same number. This sum is called the magic constant. An $n \times n$ magic square is called *normal* if it contains the number 1 through n^2. $n \times n$ magic squares can be written as linear combinations of the permutation matrices of I_n.

TABLE 2.3

Magic Square

2	7	6
9	5	1
4	3	8

The magic square in Table 2.3 is normal and has a magic constant of 15.

a. Find all of the permutation matrices of I_3

b. Are the permutation matrices that you found in part a linearly independent?

c. Defining the constant in front of the permutation matrix
$$p_6 = \begin{pmatrix} 0 & 0 & 1 \\ 0 & 1 & 0 \\ 1 & 0 & 0 \end{pmatrix}$$ to be 3, determine a linear combination of the permutation matrices in part a that generate the magic square in Figure 2.3.

d. Using your linear combination from part c, and only altering the scalar multiple of p_6, write a small for loop to determine other magic squares, which are not necessarily normal.

3

Vector Spaces

Lab 11: Vector Spaces and Subspaces

Introduction

Let V be a nonempty set of objects on which two operations are defined: addition and scalar multiplication. If the following properties hold for all u, v, and w in V and all scalars k and l, then V is a vector space.

1. (Closure under addition) If u and v are in V then $u + v$ is in V.

2. (Closure under scalar multiplication) If u is in V then ku is in V.

3. (Commutativity) $u + v = v + u$.

4. (Associativity) $u + (v + w) = (u + v) + w$.

5. (Additive identity) An additive identity, usually represented by 0, exists and is in V.

6. (Additive inverse) If u is in V then $-u$ is in V.

7. $k(u + v) = ku + kv$.

8. $(k + l)u = ku + lu$.

9. $k(lu) = (kl)u$.

10. $1u = u$.

Exercises: Let $V = R^2$ and $u = (u_1, u_2), v = (v_1, v_2)$ in V. Define addition as $u + v = (u_1 + v_1, v_2)$ and scalar multiplication as $ku = (ku_1, ku_2)$.

a. *If $u = (1,1)$ and $v = (1,2)$, find $u + v$. Is $u + v$ in V?*

b. *Under the defined addition is there an additive identity? Recall that an additive identity is a vector, v, in R^2 such that $u + v = u$, for all u in R^2.*

c. *Is V a vector space under the defined addition and scalar multiplication?*

Let $V = M_{2,2}$ where $M_{2,2}$ is all 2×2 matrices. If $u = \begin{pmatrix} u_1 & u_2 \\ u_3 & u_4 \end{pmatrix}$ and $v = \begin{pmatrix} v_1 & v_2 \\ v_3 & v_4 \end{pmatrix}$. Define addition as $u+v = \begin{pmatrix} u_1 + v_1 & u_2 + v_2 \\ u_3 + v_3 & u_4 + v_4 \end{pmatrix}$ and scalar multiplication as $ku = \begin{pmatrix} ku_1 & 0 \\ 0 & ku_3 \end{pmatrix}$.

d. If $u = \begin{pmatrix} 1 & 2 \\ 3 & 4 \end{pmatrix}$ and $k = 4$, find ku. Is ku in V?

e. What is the additive identity in V?

f. Calculate $1u$ under the defined scalar multiplication.

g. Is V a vector space under the defined addition and scalar multiplication?

h. Give an example of a set and a defined addition and scalar multiplication that is a vector space.

i. Give an example of a set and a defined addition and scalar multiplication that violates closure under scalar multiplication.

Subspaces

If W is a nonempty subset of a vector space V, then W is a subspace of V if under the operations of V

1. W is closed under addition and

2. W is closed under scalar multiplication.

Exercises:

a. Give an example of a subspace of $M_{2,2}$ under matrix multiplication and scalar multiplication.

b. Find the general solution of the homogeneous system $Ax = 0$ where $A = \begin{pmatrix} 1 & 2 \\ 2 & 4 \end{pmatrix}$. Is the set of solutions (called the solution set) to this system a subspace of R^2 under addition and scalar multiplication of vectors?

c. Let $u_1 = 1$, $u_2 = \cos(x)$, and $u_3 = \sin(x)$ be three vectors in the vector space V defined as the set of continuous functions. Is $4u_1 + 5u_2$ in V? Determine if the set of linear combinations of u_1, u_2, and u_3 is a subspace of V.

Theorems and Problems

For each of these statements, either prove that the statement is true or find a counter example that shows it is false.

Problem 39. If V is the set of 2×2 invertible matrices then V is a vector space under matrix addition and scalar multiplication of matrices.

Problem 40. The set of 2×2 symmetric matrices under matrix addition and scalar multiplication of matrices is a vector space.

Problem 41. If V is a vector space with u_1 and u_2 vectors in V then $a_1 u_1 + a_2 u_2 + b_1 u_1 + b_2 u_2 = (a_1 + b_1)u_1 + (a_2 + b_2)u_2$ for any scalars a_1, a_2, b_1, and b_2 are scalars.

Problem 42. If A is an $n \times n$ matrix, then the solution set to $Ax = 0$ is a subspace of R^n.

Problem 43. If A is an $n \times n$ matrix, then the set of linear combinations of the rows of A is a subspace of R^n.

Thereom 44. If $S = \{v_1, v_2, ..., v_n\}$ is a set of vectors in vector space V, then the set of all linear combinations of vectors in S is a subspace of V.

Lab 12: Basing It All on Just a Few Vectors

Introduction

Recall that a set S spans a vector space V if every vector in V can be written as a linear combination of vectors in S. A set S is a *basis* for a vector space V if 1) S spans V and 2) S is linearly independent.

The *dimension* of a vector space V, $dim(V)$, is the number of vectors in a basis. If a basis for a vector space, V, consists of only the 0 vector then $dim(V) = 0$. If the number of basis vectors for a vector space is finite we call the vector space *finite dimensional*, otherwise we call the vector space *infinite dimensional*.

Note that a basis for a vector space is not unique; however two different bases for the same vector space will contain the same number of vectors. In addition, if two vector spaces have the same basis then they are the same vector space.

Exercises: Let $V = R^3$.

a. *Give an example of a set of vectors in R^3 that spans V but is not linearly independent.*

b. *Give an example of a set of vectors in R^3 that is linearly independent but that does not span V.*

c. *One basis for R^3 is $S = \{(1,0,0),(0,1,0),(0,0,1)\}$, give another example of a basis for R^3.*

d. *What is the dimension of R^n?*

Nullspace

In Lab 11, we found that the general solution, *solution set*, of the homogeneous system $Ax = 0$ where $A = \begin{pmatrix} 1 & 2 \\ 2 & 4 \end{pmatrix}$, is a subspace of R^2. The solution set to $Ax = 0$ is called the *Nullspace* of A. To find a basis for the nullspace of a matrix

Type: null(The Name of the Matrix,'r')

The dimension of the nullspace of a matrix A is called the *nullity* of A. Note that if MATLAB reports that the nullspace is empty then the only vector in the nullspace is the 0 vector.

Exercises: Define $A = \begin{pmatrix} 1 & 2 \\ 2 & 4 \end{pmatrix}$, $B = \begin{pmatrix} 1 & 2 & 3 \\ 4 & 5 & 6 \\ 0 & 1 & 2 \end{pmatrix}$, *and*

$M = \begin{pmatrix} 1 & 0 & 5 & 0 \\ 0 & 1 & 3 & -1 \\ -2 & 0 & 1 & 4 \end{pmatrix}$.

a. Find a basis for the nullspace of A and the nullity of A.

b. Find a basis for the nullspace of B and the nullity of B.

c. From parts a and b make a conjecture about the nullspace and nullity of invertible matrices.

d. Compare the nullity of A with the nullity of A^T and the nullity of B with the nullity of B^T.

e. Note that matrix M is not square but we can still find a basis for the nullspace of M, so find a basis for the nullspace of M and the nullity of M.

Rowspace and Columnspace

The *rowspace* of A is the set of vectors that can be written as linear combinations of the rows of A. The dimension of the rowspace of A is the *rank* of A, $rank(A)$. Similarly the *columnspace* of A is the set of vectors that can be written as linear combinations of the columns of A. The dimension of the columnspace of A is also the $rank(A)$.

The vectors of one basis for the rowspace of A are the nonzero rows of A in reduced row echelon form. Unlike the rows of A, the columns of A are affected by row operations. So to find the vectors of a basis for the columnspace of A, you can put A in reduced row echelon form and then identify the columns with the leading ones. The corresponding columns in the original A will create a basis for the columnspace of A.

Exercises: Using matrices A, B, and M from above

a. Find a basis for the rowspace of A and $rank(A)$.

b. Determine $rank(A) + nullity(A)$.

c. Determine if the basis for the rowspace of A from part a spans R^2.

d. If an $n \times n$ matrix is invertible make a conjecture about the relationship between the rowspace of the matrix and R^n.

e. Would a similar result to that in d hold for the columnspace of an $n \times n$ invertible matrix? If you are unsure try finding the columnspace of A.

e. *Find a basis for the rowspace and columnspace of* M.

f. *Determine* $rank(M) + nullity(M)$.

g. *Make a conjecture about the sum of the rank and nullity of a square matrix. What is the sum of the rank and nullity of a matrix in general?*

Theorems and Problems

For each of these statements, either prove that the statement is true or find a counter example that shows it is false.

Theorem 45. A is invertible if and only if the nullspace of $A = \vec{0}$ and $nullity(A) = 0$.
Theorem 46. An $n \times n$ matrix A is invertible if and only if the rowspace of $A = R^n$ and $rank(A) = n$.
Problem 47. If A is an $m \times n$ matrix then $rank(A) + nullity(A) = m$.
Problem 48. $rank(A) = rank(A^T)$.

Now how many different statements can you think of that are equivalent to saying that a square matrix, A, is invertible?

If A is an $n \times n$ matrix the following are equivalent statements:

1. A is invertible.

2. $|A| \neq 0$.

3. The reduced row echelon form of A is I_n.

4. A can be written as a product of elementary matrices.

5. The system $Ax = b$ has exactly one solution for all $n \times 1$ vectors b.

6. The system $Ax = 0$ has only the trivial solution.

7. The nullspace of $A = \vec{0}$ and $nullity(A) = 0$.

8. The rowspace of $A = R^n$, the columnspace of $A = R^n$, and $rank(A) = n$.

Lab 13: Linear Transformations

Introduction

A transformation $T : V \longrightarrow W$ is a mapping between vector spaces V and W. The transformation is a *linear transformation* if and only if $T(\vec{0}) = \vec{0}$, $T(v_1 + v_2) = T(v_1) + T(v_2)$, and $T(kv_1) = kT(v_1)$ for all vectors v_1 and v_2 in V and scalar k.

We will be working with transformations of the form $T(x) = Ax$. We call the matrix A the *standard matrix*.

Basic Linear Transformations and Standard Matrices

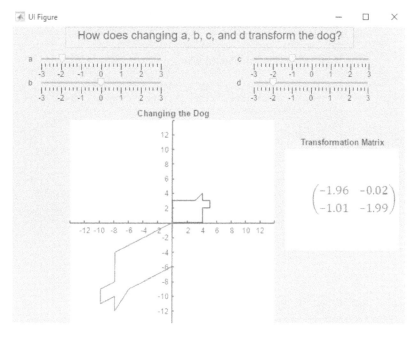

FIGURE 3.1

Exercises:

a. *Change the values of a and/or d in* https://www.mathworks.com/ *matlabcentral/fileexchange/64916-transforming-the-dog to reflect the dog over the x axis, y axis, and then the origin. What are the standard matrices for each of these three transformations?*

b. *Change the values of a and d, leaving $b = c = 0$, in the demonstration to* stretch the dog in both the x and y direction. A dilation is when the dog is stretched and a contraction is when the dog is shrunk. What values of a and d relate to a dilation of the dog?

c. What would the standard matrix look like if you wanted to reflect the dog over the line $y = x$?

d. A projection onto an axis or a line is done by dropping a perpendicular line segment from each point on the image to the line that you are projecting onto. What should the dog look like if you project onto the x axis? What about if you project onto the y axis? Use the demonstration to determine what standard matrices produce these images.

e. Use the demonstration $https://www.mathworks.com/matlabcentral/$ $fileexchange/64917-transforming-the-dog-with-rotation$ to determine the standard matrix affiliated with rotating counterclockwise 45 degrees.

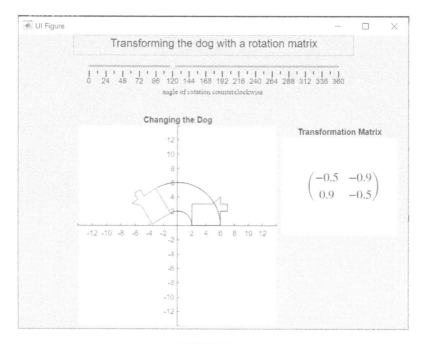

FIGURE 3.2

f. The standard matrix affiliated with composition of transformations $T_A \circ T_B(x)$ is AB. Calculate the standard matrix affiliated with the following sequence of transformations 1) Reflect over $y=x$, 2) Rotating counterclockwise 45 degrees, and 3) dilating (scaling factor) by a factor of 3/2.

Use the demonstration $https://www.mathworks.com/matlabcentral/$
$fileexchange/66107-transforming-the-dog-with-a-composition$
$-of-linear-transformations$ *to visualize the composition. Note that the
order that transformations are performed may matter.*

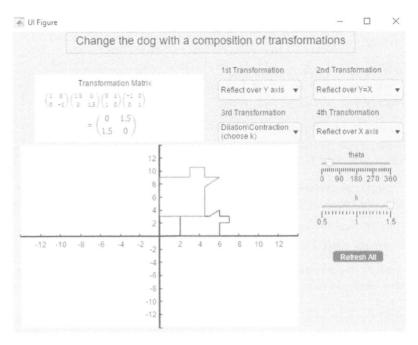

FIGURE 3.3

One-to-One Transformations

A transformation is *one-to-one* if $T(v_1) = T(v_2)$ implies $v_1 = v_2$.

*Exercise: Think about the transformations that you explored above, rotation,
reflection, projection, dilation, and contraction; which transformations are one
to one? How is the determinant of the standard matrix related to whether the
transformation is one-to-one?*

Transformations from $R^n \to R^m$

So far we have only seen a few special transformations from $R^2 \to R^2$. We
can also explore other transformations.

The *kernel* of transformation T is the set of vectors that T maps to $\vec{0}$.

That is x is in the kernel of T if $T(x) = \vec{0}$. The dimension of the kernel is called the *nullity* of T, *nullity*(T).

If $T : V \to W$ then the *range* of transformation T is the set of vectors y in W such that $T(x) = y$ for some x in V. The dimension of the range of T is called the *rank* of T, *rank*(T).

Exercises:

a. *Define the transformation $T : R^2 \to R^2$ as $T(x_1, x_2) = (x_1 - x_2, 2x_1 + x_2)$. Find a basis for kernel of T, nullity(T), a basis for the range of T and rank(T).*

b. *Define the transformation $T_A : R^3 \to R^3$ as $T_A(x_1, x_2, x_3) = (x_1 + x_2, 2x_2, x_1 - x_3)$. What is the standard matrix affiliated with T_A? Is T_A a one-to-one transformation?*

c. *Describe the relationship between kernel of T_A and the nullspace of A and between range of T_A and the columnspace of A from part b.*

Theorems and Problems

For each of these statements, either prove that the statement is true or find a counter example that shows it is false.

In all of the following statements, $T_A : R^n \to R^n$ is defined by multiplication by the $n \times n$ standard matrix A.

Theorem 49. T_A is one-to-one if and only if A is invertible.
Theorem 50. If A is an invertible matrix then the kernel of $T_A = R^n$.
Theorem 51. If A is an invertible matrix then $nullity(T_A) = \vec{0}$.
Problem 52. If $T_1 : R^n \to R^m$ and $T_2 : R^m \to R^p$ are two linear transformations then $T_2 \circ T_1$ is a linear transformation.
Problem 53. $T_1(x_1, x_2) = (x_1 + k_1, x_2 + k_2)$ is a linear transformation where k_1 and k_2 are nonzero scalars. (This transformation represents translation.)

Lab 14: Eigenvalues and Eigenspaces

Introduction

Recall that given a square matrix A we can calculate the eigenvalues of A by finding the values for λ such that $Ax = \lambda x$. The λ's are the eigenvalues for A and each eigenvalue has a corresponding eigenvector x.

Given λ, its corresponding eigenvector, x, can be found by solving for x in $Ax = \lambda x$. Note that in fact there will be infinitely many solutions to this system and thus we can discuss the eigenvectors in terms of a basis for the eigensystem corresponding to λ.

Another way to find the eigenvalues is to solve for λ in the characteristic equation $|A - \lambda I| = 0$.

An $n \times n$ matrix has n eigenvalues (counting algebraic multiplicity). The algebraic multiplicity of λ is its multiplicity as a root of the characteristic polynomial. The geometric multiplicity of an eigenvalue λ is the dimension of the eigenspace associated to λ.

To find the eigenvalues of a matrix,

Type **eig(The Name of the Matrix)**.

If you type

[V,E]=eig(The Name of the Matrix)

the diagonal entries of E are the eigenvalues of the matrix and columns of matrix V are the corresponding eigenvectors.

Exercises: Let $A = \begin{pmatrix} 1 & 2 & 3 \\ 0 & 5 & 6 \\ 0 & 0 & 0 \end{pmatrix}$, $B = \begin{pmatrix} 1 & 2 & 3 \\ 0 & 5 & 6 \\ 0 & 0 & 5 \end{pmatrix}$, *and*

$M = \begin{pmatrix} 1 & 2 \\ 2 & 4 \end{pmatrix}$.

a. *Find the eigenvalues of A, B, and M.*

b. *Make a conjecture about the values of the eigenvalues of singular matrices.*

c. *Find a basis for the eigenspace for each of the eigenvalues for matrices B and M. Note that a basis for the eigenspace affiliated with an eigenvalue consists of the eigenvectors affiliated with that eigenvalue.*

d. *Are the eigenvectors of M linearly independent? Explain your answer.*

e. *Compare the sum of the eigenvalues of M with the trace of M. How are they related?*

f. *Compare the product of the eigenvalues of M with the determinant of M. How are they related?*

g. *The eigenvectors of B are linear dependent. Make a conjecture about the property of B that causes this to be true.*

h. *To find the coefficients, $p_1, p_2, \ldots p_n, p_{n+1}$ of the characteristic polynomial $p_1 x^n + p_2 x^{n-1} + \cdots p_n x + p_{n+1}$ for a $n \times n$ matrix type:*

poly(Name of the Matrix).

An $n \times n$ matrix will have an nth degree characteristic polynomial associated with it. Find the characteristic polynomial for B.

i. *The roots of the characteristic polynomial for a matrix are the eigenvalues of the matrix. Use the characteristic polynomial from h. to find the eigenvalues of B.*

j. *If $p(x)$ is the characteristic polynomial in h, $p(x) = 0$ is the characteristic equation, determine $p(B)$ keeping in mind that if there is a constant term, k, in the characteristic polynomial, $p(x)$, that term is kI in $p(B)$. What property do you notice when computing $p(B)$?*

Cayley–Hamilton Theorem

In part j above you may have noticed that $p(B) = 0$, where 0 is the zero-matrix. In general, if A is an $n \times n$ matrix and $p(x)$ is the characteristic polynomial for A, then $p(A)$ results in the zero-matrix.

You may recall seeing the Cayley–Hamilton Theorem in Lab 5 relate the trace of a matrix with its determinant. Here we will see how it can be used to find the inverse of a matrix. If A is an $n \times n$ invertible matrix then

$$A^{-1} = \frac{(-1)^{n-1}}{det(A)} (A^{n-1} + p_2 A^{n-2} + p_3 A^{n-3} \cdots + p_n I)$$

where the characteristic polynomial of $A = A^n + p_2 A^{n-1} + \cdots + p_n A + (-1)^n det(A) I$.

Exercise: Use the Cayley–Hamilton Theorem to find B^{-1}.

Theorems and Problems

For each of these statements, either prove that the statement is true or find a counter example that shows it is false.

Theorem 54. A is invertible if and only if all its eigenvalues are nonzero.

Problem 55. If λ is an eigenvalue for A then $\frac{1}{\lambda}$ is an eigenvalue for A^{-1}.

Problem 56. If λ is an eigenvalue for A then λ^k is an eigenvalue for A^k.

Problem 57. If λ is an eigenvalue for A then λ is an eigenvalue for A^T.

Problem 58. The eigenvalues of a triangular matrix (upper or lower) are the entries on the main diagonal.

Problem 59. If A is an $n \times n$ matrix with n distinct eigenvalues then all of the corresponding eigenvectors are linearly independent.

Lab 15: Markov Chains: An Application of Eigenvalues

Introduction

This is a discrete modeling technique for modeling systems that undergo transitions between a finite (or countable) number of states. Each Markov chain has a corresponding *transition matrix*. The transition matrix, M, is a probability matrix, where $M_{i,j}$ is the probability of going from state j to state i. That is, M is of the form

$$
\begin{array}{c}
\text{NewState 1} \\
2 \\
3
\end{array}
\begin{pmatrix}
\text{Preceding State} & & \\
1 & 2 & 3 \\
0.05 & 0.7 & 0.46 \\
0.75 & 0.2 & 0.12 \\
0.2 & 0.1 & 0.42
\end{pmatrix}.
$$

It is sometimes helpful in a Markov chain to visualize the system with a state diagram, with arrows between states representing the transitions (and weights representing the probability of that transition).

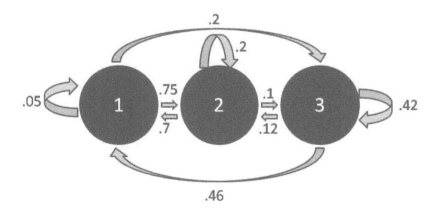

FIGURE 3.4: Example of a state diagram

In addition to the transition matrix, each Markov chain has an initial vector which is a vector typically consisting of initial total populations in each state or a fraction of the total population in each state.

Exercises:

a. *Type the MATLAB commands below,*

$$M = [.05 \ .7 \ .46; .75 \ .2 \ .12; .2 \ .1 \ .42];$$
$$x0 = [100; 0; 0];$$
$$loops = 40;$$
$$for \ k = 1 : loops$$
$$\quad bar(mpower(M,k) * x0);$$
$$\quad drawnow;$$
$$\quad pause(.1);$$
$$\quad F(k) = getframe(gcf);$$
$$end$$

The movie shows what is happening in the 3 states as k increases from 1 to 40. Identify the transition matrix, M, and the initial vector in this problem.

b. *Describe what Mx_0 represent. What does $M^k x_0$ represent?*

c. *For what value of k does the system appear to become stable?*

d. *Type and evaluate the MATLAB commands below which shows only the population of state 1 after k steps. For what value of k does this state's population appear to stabilize?*

$$clear \ x; clear \ y;$$
$$M = [.05 \ .7 \ .46; .75 \ .2 \ .12; .2 \ .1 \ .42];$$
$$x0 = [100; 0; 0];$$
$$for \ i = 1 : 40$$
$$x = linspace(0,i,i+1);$$
$$for \ k = 1 : i+1$$
$$\quad temp = mpower(M,k) * x0;$$
$$\quad y(k) = temp(1);$$
$$end;$$
$$\quad scatter(x,y);$$
$$\quad drawnow;$$
$$\quad pause(.1);$$
$$F(k) = getframe(gcf);$$
$$end$$

The transition matrix has a dominant eigenvalue, which is the largest eigenvalue in magnitude. A Markov chain has a stable solution if the dominant eigenvalue has a magnitude of 1.

Exercises: The inhabitants of a vegetarian-prone community agree on the following rules

1. *Only one out of six people will eat meat the next day if they eat meat today.*

2. *A person who eats no meat one day will flip a fair coin and eat meat on the next day if and only if a head appears.*

If 80% of the population eat meat on the first day, in the long run, what percentage of the population will eat meat each day?

a. *Construct the transition matrix for this problem.*

b. *If 80% of the population eat meat on the first day, then the initial population vector is (.8,.2). Graph the percent of the population that will eat meat versus time for the first 10 days and interpret this graph. (Hint: Use scatter to graph individual points, (time, meat eating population %.))*

c. *Find the steady state vector for this problem using the graph and interpret your results.*

d. *Find the eigenvalues and eigenvectors of the transition matrix.*

e. *Using the eigenvector corresponding to the eigenvalue of magnitude one, create a percentile vector. In order to make this vector a percentile state vector, the total of the two values should equal one. What scalar do you need to multiply this vector by in order to make it a percentile state vector? What does this vector represent?*

Project Set 3

Project 1: Computer Graphics

The purpose of this exercise is to introduce you to the topic of linear transformation as they relate to computer graphics.

a. Create a list of points that when attached will create a block letter graphic representing your first initial. Graph the letter you created. An example is below.
$T = [3\ 3\ 1\ 1\ 6\ 6\ 4\ 4\ 3; 1\ 6\ 6\ 7\ 7\ 6\ 6\ 1\ 1]$;
$x = T(1, :)$;
$y = T(2, :)$;
$plot(x, y)$;

b. Create a standard matrix that would transform your original graphic into a graphic that is 4 times larger along the x axis and $1/2$ as large along the y axis. Graph the transformed graphic and make sure to use your standard matrix in your solution.

c. Create a translation that would move your original graphic 6 to the right and 3 units up. Graph the transformed graphic.

d. Create a standard matrix that will reflect your original graphic about the origin. Graph the transformed graphic and make sure to use your standard matrix in the solution.

e. Create a sequence of transformations that will reflect the original graphic over the line $y = 6$. Graph the transformed graphic.

f. Create a sequence of transformations that will reflect your original graphic about the point (2,3). Graph the transformed graphic.

Project 2: Fractals

A fractal is an iterative system defined by a set of rules. In this project, you will start with F_0 as the polygon and the rule is $F_i = (F_i)^1 \cup (F_i)^2 \cup (F_i)^3$ where $(F_i)^1 = A(F_{i-1}) + b_1$, $(F_i)^2 = A(F_{i-1}) + b_2$, $(F_i)^3 = A(F_{i-1}) + b_3$, where A is a contraction matrix and b_1, b_2, and b_3 are translations.

The program below generates the first two steps in Sierpinski's Triangle. Alter the program, integrating a for loop, to generate the first 8 iterations (pictures shown in Figure 3.5).

FIGURE 3.5: The first 8 iterations of Sierpinski's Triangle

$A = [1/2\ 0; 0\ 1/2]$;
$b_1 = [1/4\ 1/4\ 1/4\ 1/4; 0\ 0\ 0\ 0]$;
$b_2 = [3/4\ 3/4\ 3/4\ 3/4; 0\ 0\ 0\ 0]$;
$b_3 = [1/2\ 1/2\ 1/2\ 1/2; 1/2\ 1/2\ 1/2\ 1/2]$;
$Triangle = [1/2\ 1\ 3/2\ 1/2; 0\ 1\ 0\ 0]$;
$x = Triangle(1, :); y = Triangle(2, :)$;
$fill(x,y,'black')$
$Block1 = A * Triangle + b_1$;
$xb1 = Block1(1, :); yb1 = Block1(2, :)$;
$Block2 = A * Triangle + b_2$;
$xb2 = Block2(1, :); yb2 = Block2(2, :)$;
$Block3 = A * Triangle + b_3$;
$xb3 = Block3(1, :); yb3 = Block3(2, :)$;
$figure$
$hold\ on$
$fill(xb1,yb1,'black')$
$fill(xb2,yb2,'black')$
$fill(xb3,yb3,'black')$
$hold\ off$

Project 3: Genetics

A certain trait is determined by a specific pair of genes, each of which may be two types, say R or r. An individual may have:

1. RR combination (dominant)

2. Rr or rR, considered equivalent genetically (hybrid)

3. rr combination (recessive)

Offspring inherit one gene of the pair from each parent. Genes inherited from each parent are selected at random, independently of each other. This determines probability of occurrence of each type of offspring.

In this project, we will be looking at tongue rolling. Tongue rolling, the ability to roll the tongue, is a dominant trait (R), while non-rolling is recessive (r). At each generation someone of unknown genetic makeup mates with a hybrid.

So the possibilities at each generation are RR, Rr, and rr. In the community of Lolly, 50% of the current generation, 0th generation, are RR and 50% are Rr.

a. If at the 0th generation, the parents are RR and Rr what is the probability that the offspring is Rr?

b. What would the transition matrix look like from the previous generation to the next generation? What would the initial vector look like?

c. What would be the percent of RR, Rr, and rr in the next generation?

d. What percentage of the 4th generation in Lolly are in each state (genetic makeup for tongue rolling)?

e. In the long run, many generations, what will the percent of people in each state be in Lolly? How could we have determined this through inspection using eigenvalues and eigenvectors of the transition matrix?

f. Write up your findings and supporting mathematical argument.

Project 4: Tree Harvesting

Sixty-one percent of the state of North Carolina is forestland. Loblolly pine is the most important commercial timber in the southeastern United States. Over 50% of the standing pine in the southeast is loblolly. This is an easily seeded, fast-growing member of the yellow pine group. On an average site, the loblolly would reach 55-65 feet in 25 years. Thinning of loblolly pine farms should start around 15-20 years.

The goal of this problem is to determine the number of trees to harvest. Let's say that we have planted loblolly pines in our plantation for the past 15 years and thus there are trees at a variety of heights, which we will put into categories, $p_1, p_2, \ldots p_n$. After 15 years we wish to thin our plantation and thus will harvest trees from each category. The matrix that represents the growth rates is called the growth matrix and is of the form

$$G = \begin{pmatrix} 1-g_1 & 0 & 0 & \cdots & 0 & 0 \\ g_1 & 1-g_2 & 0 & \cdots & 0 & 0 \\ 0 & g_2 & 1-g_3 & 0 & 0 & 0 \\ \vdots & \vdots & \ddots & \ddots & \vdots & \vdots \\ 0 & 0 & \cdots & 0 & 1-g_{n-1} & 0 \\ 0 & 0 & \cdots & 0 & g_n & 1 \end{pmatrix}.$$

a. At the 15-year marker, the beginning of harvesting,

$$x = \begin{pmatrix} x_1 \\ x_2 \\ x_3 \\ x_4 \\ x_5 \end{pmatrix}$$ represents the number of trees in each category. Assuming

the growth is calculated such that the growth matrix G transitioning from one year to the next is

$$G = \begin{pmatrix} 1 & 0 & 0 & 0 & 0 \\ 0.75 & 0.4 & 0 & 0 & 0 \\ 0 & 0.6 & 0.5 & 0 & 0 \\ 0 & 0 & 0.5 & 0.6 & 0 \\ 0 & 0 & 0 & 0.4 & 1 \end{pmatrix},$$

what does Gx represent? From the matrix G you might note that 75% of trees in category 1 move to category 2 in a year (time period), what might the farmer be doing to make the (1,1) entry of G equal to 1?

b. Suppose h_i is the fraction of the ith category that will be harvested at the end of each year, and we let H be the diagonal matrix whose entries are the h_i's. What does HGx represent? What does $Gx - HGx$ represent?

c. Assume that $x_1 = 100$, if $H = \begin{pmatrix} 0 & 0 & 0 & 0 & 0 \\ 0 & 0.1 & 0 & 0 & 0 \\ 0 & 0 & 0.1 & 0 & 0 \\ 0 & 0 & 0 & 0.2 & 0 \\ 0 & 0 & 0 & 0 & 0.8 \end{pmatrix}$, use G and H

to determine how to maintain a sustainable tree farm. What does the 0 in the (1,1) entry of H represent?

d. Describe how your solution in c. is related to the concept of eigenvalues and eigenvectors.

Project 5: Sports Ranking

In Project Set 1, we looked at ranking the teams in the Big Ten using powers of matrices.

Michigan State W – Indiana	Michigan State W – Purdue
Michigan State W – Illinois	Michigan State W – Iowa
Indiana W – Penn State	Penn State W – Michigan
Iowa W – Minnesota	Iowa W – Northwestern
Michigan W – Minnesota	Michigan W – Indiana
Minnesota W – Northwestern	Minnesota W – Wisconsin
Minnesota W – Nebraska	Nebraska W – Purdue
Nebraska W – Illinois	Ohio State W– Wisconsin
Ohio State W – Penn State	Ohio State W – Iowa
Ohio State W – Northwestern	Wisconsin W – Illinois
Wisconsin W – Northwestern	Wisconsin W – Purdue

In this project, will be working with a *preference matrix*, A, where $a_{i,j} = w_{i,j}/n_i$, $w_{i,j}$ is the number of times team i beats team j and n_i is the number of games played by team i.

a. Create the preference matrix, A, for the Big Ten games played.

b. Determine the ranking vector r such that $Ar = \lambda r$, where λ is the eigenvalue of largest magnitude. In this ranking, the strength of a team is proportional to its score.

c. Discuss how you might integrate the strength of schedule into the matrix A and explain why you believe this will better the ranking.

Similar techniques to these are used in the Google Page Rank and other searches with weights given to links instead of wins.

Project 6: Seriation and the Fiedler Vector

In Project Set 1, we introduced the idea of seriation applied to archeology, where we ordered artifacts based on minimizing dissimilarities. This technique required looking at m permutations of the original artifact-trait matrix, where m is the number of artifacts. Other techniques must be explored if the number of artifacts is larger than 12.

The technique presented here guarantees a minimum ordering only if a permutation matrix can be found that when applied to the artifact-trait matrix eliminates all of the embedded zeros. This is not very practical, but the technique does a decent job of ordering even if not all embedded zeros can be removed.

Given an $m \times m$ symmetric matrix S and a diagonal matrix D such that $D_{i,i} = \sum_{j=1}^{m} S_{i,j}$ for $1 \le i \le m$, the *Laplacian matrix* of S is $L = D - S$. The eigenvector associated with the second smallest eigenvalue of L, the *Fiedler value*, is the Fiedler vector. The permutation which puts the the Fiedler vector in increasing order is the ordering of the artifacts.

In this example, we wish to order pieces of music, presented in Figure 3.6, based on their traits. The goal is to determine which musical pieces are most similar based on these traits.

SONG	Crescendo	Decrescendo	Staccato	Portamento	Tempo	Intervals
Beat It	4	3	18	0	132	2.417
How Will I Know	17	4	34	0	120	5.281
Together Again	4	2	16	0	120	5.875
No Scrubs	10	4	14	24	90	5.563
Poker Face	4	2	16	8	120	4.75
Crazy in Love	7	4	19	24	100	5.563
Only Girl	18	2	56	11	126	3.24
Say My Name	10	2	21	26	70	5.25
Dreamlover	6	4	19	8	104	5.375
Like a Virgin	4	0	14	0	118	2.938
Billie Jean	9	4	14	9	117	3.15
I Wanna Dance	10	2	38	12	120	1.563
All for You	0	4	10	15	114	4.389
Waterfalls	3	2	15	8	84	1.5
Born This Way	9	1	14	9	124	2.656
Single Ladies	8	0	36	28	96	5.292
Umbrella	5	6	34	4	86	4.375
Bootylicious	14	2	20	16	104	7.75
Fantasy	7	1	16	7	112	5.375
Like a Prayer	3	1	10	0	120	4.375

FIGURE 3.6: Raw data for 20 #1 Billboard Hit choruses

a. This technique requires a binary matrix (a matrix of zeros and ones). Use the raw data from Figure 3.6 to create a song-trait matrix. In order to do so, use the following values to determine whether to assign a zero or a one to each raw data value. In the crescendo category, assign a 0 for raw data less than 8 and assign a 1 otherwise, and we'll call 8 the cut off for this category. Use cutoffs of 2 for decrescendos, 22 for staccato, 10 for portamento, 109 for tempo, and 4 for intervals.

b. Find the similarity matrix, S, related to the binary song-trait matrix from a (see Project Set 1 for more information on similarity matrices).

c. Let D be a diagonal matrix with diagonal entries $D_{i,i} = \sum_{j=1}^{m} S_{i,j}$ for $1 \le i \le m$ using the matrix S from part b. Find the Laplacian matrix, $L = D - S$ and the eigensystem affiliated with L.

d. Using the Laplacian matrix from part c, determine the Fiedler value, the Fiedler vector, and the ordering of the musical pieces.

e. Those musical pieces that are close together in the ordering are most similar in their traits. Interpret your results. Are there songs from the same artist close together in the ordering? Are there other traits that you would add to the study to get stronger results if you furthered the study?

Project 7: Hamming Codes

With each of our daily lives today filled with the need for technology, the need for safe and accurate data transmission is essential. In this project, we will discuss a way to detect whether a binary message has been altered from its original state through the transmission process and possible addition of noise. Noise in a binary message may make a value of 0 into a 1 or visa versa.

One way to check if the message has changed is for the sender to add a *single parity check bit* to the end of the message. This single bit would be a binary number which would make the full message have even parity.

Example If a 4-bit message is 1011 then the sent message with the single parity bit would be the 5-bit message 10111 since the sum of the digits is $4 \equiv 0(mod2)$ and thus the number has even parity.

In this project, we will be working with *Hamming Code* error correction which finds and corrects a single error transmission using multiple parity check bits. As you can see from the example above, parity check bits are appended to the end of the message. The set of binary messages with their appended parity check bits form a vector space, H_5, under modulo-2 addition and scalar multiplication. Using our example above, 10111 is a vector in the vector space of 5-bit messages under modulo-2 addition and scalar multiplication.

a. Add the two vectors 10111 and 10010 in H_5.

b. Determine a basis for H_5 and the dimension of H_5.

Hamming Codes require the addition of 3 parity check bits in order to correct a single transmission error. Let x_1, x_2, x_3, and x_4 be the 4 binary values in the original message and x_5, x_6, and x_7 be the 3 parity check bit values, where

$$\begin{aligned} x_1 + x_2 + x_4 + x_5 &\equiv 0(mod2), \\ x_1 + x_3 + x_4 + x_6 &\equiv 0(mod2), \\ x_2 + x_3 + x_4 + x_7 &\equiv 0(mod2). \end{aligned}$$

For any received message m_r, the product Am_r is called the *syndrome vector*.

c. Write the 4 bit message 1011 with the three parity changes (thus a 7-bit message).

d. Write the above equations as a homogeneous system $A\vec{x} = \vec{0}$. A is called the *parity check matrix*.

e. Denote the set of codes $\vec{x}$, or *code space*, C_4; find a basis for C_4. What is the dimension of C_4?

f. The parity check matrix, A, will help check to determine if the message was received correctly. If a message, m_r, is received correctly then $Am_r \equiv \vec{0}(mod2)$. If the message you received is $m_r = \{0,1,1,1,0,0,1\}$ was it received correctly? How do we know that the message $m_r = \{0,0,1,0,1,1,0\}$ was transmitted incorrectly?

g. In order to detect and correct the message $m_r = \{0,0,1,0,1,1,0\}$ that was transmitted incorrectly, inspect the syndrome vector and determine which equations, from the homogeneous system, have an error. This can be done by identifying which entries in the syndrome vector are nonzero modulo 2. (Recall a correct transmission will produce all zeros modulo 2 in the syndrome vector.)

h. **Example:** If Equation 1, $x_1 + x_2 + x_4 + x_5 \equiv 0(mod2)$, is incorrect then either x_1, x_2, or x_4 is incorrect. If Equation 1 is correct then x_1, x_2, and x_4 are correct.

Using your results from part g determine which single bit, x_1, x_2, x_3, or x_4, is incorrect.

4

Orthogonality

Lab 16: Inner Product Spaces

Introduction

An *inner product* on set V is a function that maps ordered pairs (x,y) from $V \times V$ (that is x and y are elements of V) to a number $< x,y >$ while satisfying the following properties:

1. For all v in V, $< v,v > \geq 0$ and $< v,v >= 0$ if and only if $v = \vec{0}$.

2. For all u, v, and w in V, $< u,v + w >=< u,v > + < u,w >$.

3. For all u and v in V and scalar k,

$$< ku,v >=< u,kv >= k < u,v > .$$

4. For all u and v in V, $< u,v >= \overline{< v,u >}$.

A vector space with a defined inner product is called an *inner product space*.

The *Euclidean Inner Product* in R^n is defined as

$$< u,v >= u_1 v_1 + u_2 v_2 + u_3 v_3 + \cdots + u_n v_n$$

where $u = (u_1, u_2, u_3, \cdots, u_n)$ and $v = (v_1, v_2, v_3, \cdots, v_n)$ are in R^n.

Use the Euclidean inner product in R^2, also called the *dot product*, for the following exercises. To calculate the Euclidean inner product, $< u,v >$, type: **dot(u,v)**

Exercises: Let $u = (3,2)$, $v = (2,0)$, and $w = (0,2)$ be vectors in R^2.

a. *Determine $< u,v >$.*

b. *The length, also called the norm or magnitude, of a vector u is $||u|| = \sqrt{< u,u >}$. To calculate $||u||$, type **norm(u)**. Determine $||u||$.*

c. *A vector is said to be a unit vector if its magnitude is 1. Are u, v, or w unit vectors? What scalar can you multiply vector v by to make it a unit vector?*

73

d. *We think of orthogonality as perpendicularity in R^2. Are v and w orthogonal? In general, two vectors u and v are orthogonal if $< u,v >= 0$. Are v and w orthogonal using this definition?*

More about Orthogonality

A set is called *orthogonal* if each pair of vectors in the set is orthogonal. If, in addition, every vector in the set has a magnitude of 1, then the set is called *orthonormal*. An *orthonormal basis* is a basis for a vector space which is orthonormal.

If V_1 is a subset of V, then the *orthogonal complement* of V_1 is the set of all vectors in V that are orthogonal to every vector in V_1.

Exercises:

a. *Give an example of an orthonormal basis of R^3.*

b. *For any $U = \begin{pmatrix} u_1 & u_2 \\ u_3 & u_4 \end{pmatrix}$ and $V = \begin{pmatrix} v_1 & v_2 \\ v_3 & v_4 \end{pmatrix}$ in $M_{2,2}$, define the inner product as $< U,V >= u_1v_1 + u_2v_2 + u_3v_3 + u_4v_4$. Find an orthonormal basis for $M_{2,2}$ under this inner product.*

c. *Find a basis for the nullspace of $A = \begin{pmatrix} 1 & 2 & 3 \\ 4 & 5 & 6 \\ 7 & 8 & 9 \end{pmatrix}$ and a basis for the rowspace or A. Are the nullspace of A and the rowspace of A orthogonal complements?*

d. *Determine if the nullspace of A^T and the columnspace of A are orthogonal complements.*

Gram–Schmidt Process

The *Gram–Schmidt Process* is a numerical technique for finding an orthonormal set of vectors that spans an inner product space, usually R^n.

The algorithm in R^n

1. Start with n linearly independent vectors $\{v_1, v_2, \cdots, v_n\}$.

2. Let $u_1 = v_1$ and $e_1 = \frac{u_1}{||u_1||}$.

3. Project v_2 onto u_1, $proj_{u_1}v_2 = \frac{<u_1,v_2>}{||u_1||^2}u_1$.

4. Define $u_2 = v_2 - proj_{u_1}v_2$ and $e_2 = \frac{u_2}{||u_2||}$.

5. In general, $u_n = v_n - \sum_{i=1}^{n-1} proj_{u_i}v_n$ and $e_n = \frac{u_n}{||u_n||}$.

Exercise: $\{v_1 = (1,2,2), \ v_2 = (4,5,6), \ v_3 = (8,9,1)\}$ *is a set of linearly independent vectors in* R^3. *Use the Gram–Schmidt process and* v_1, v_2, *and* v_3 *to find an orthonormal set of vectors* $\{e_1, e_2, e_3\}$ *that span* R^3 *(Keep in mind that there may be some round off error here so if your dot products are very small but not 0 in MATLAB, you can assume that they are 0).*

Theorems and Problems

For each of these statements, either prove that the statement is true or find a counter example that shows it is false.

Theorem 60. If u and v are orthogonal vectors in V then

$$||u + v||^2 = ||u||^2 + ||v||^2.$$

Theorem 61. If u and v are vectors in V then

$$| < u,v > | \leq ||u||||v||.$$

Theorem 62. If u and v are vectors in V then $||u + v|| \leq ||u|| + ||v||$.
Problem 63. If u and v are vectors in V then

$$||u + v||^2 + ||u - v||^2 = 2(||u||^2 + ||v||^2).$$

Theorem 64. If $\{v_1, v_2, \cdots, v_n\}$ is an orthonormal set of vectors in V, then $||k_1 v_1 + k_2 v_2 + \cdots k_n v_n||^2 = |k_1|^2 + |k_2|^2 + \cdots + |k_n|^2$.
Theorem 65. Every orthonormal set of vectors is linearly independent.
Problem 66. If $T : R^n \to R^n$ is the linear transformation defined by left multiplication of the $n \times n$ matrix A and vector v is in the range of T then v is orthogonal to the nullspace of A.

Lab 17: The Geometry of Vector and Inner Product Spaces

Triangle Inequality

There are many properties of inner product spaces that we will explore in this lab. We begin by exploring the Triangle Inequality.

Exercises: If $V = R^2$ with the Euclidean inner product,

a. *Use the demonstration* $https://www.mathworks.com/matlabcentral/$ $fileexchange/64926-sum-of-two-vectors$. *If u and v are the main vectors in this demonstration, represented in red and blue, choose several values for the x and y coordinates of these vectors, and note the value of both $||u|| + ||v||$ and $||u + v||$, the magnitude of the purple vector. Which quantity is always larger?*

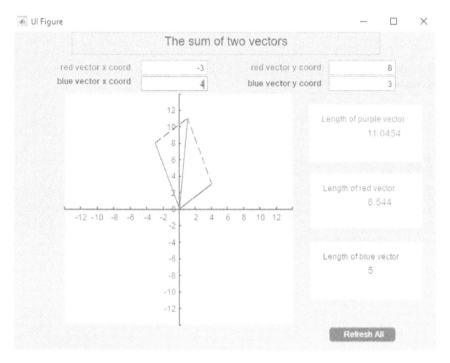

FIGURE 4.1

b. *Determine vectors u and v such that $||u|| + ||v|| = ||u + v||$.*

c. *If V is the vector space of continuous functions on $[-1,1]$ and $< f, g >=$ $\int_{-1}^{1} f(x)g(x)dx$, let $f(x) = x^2$ and $g(x) = x$ compute $||f(x)||, ||g(x)||$ and*

$||f(x) + g(x)||$. *To find the integral of a function,* $f(x)$, *on the interval* $[a,b]$ *define the function* fun, *for example* **fun** $=$ **@(x)x.^2** *defines the function as* x^2 *and then type* ***integral(fun,a,b).***

d. *Do you find the same property using the inner product from part c as you did in part a? To explore the triangle inequality with functions further, use the demonstration* $https: // www. mathworks. com/ matlabcentral/ fileexchange/64935-triangle-inequality-with-functions.$

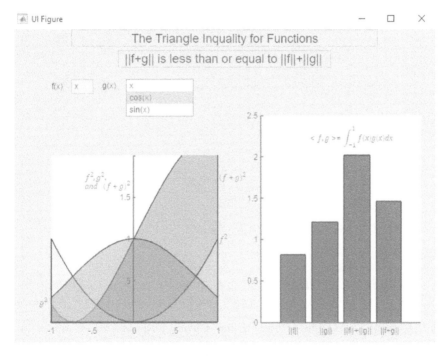

FIGURE 4.2

In general, the Triangle Inequality state that for all vectors u and v in vector space V,

$$||u + v|| \leq ||u|| + ||v||.$$

Cauchy–Schwarz Inequality

Exercises: If $V = R^2$ *with the Euclidean inner product,*

a. *Use the demonstration* $https: // www. mathworks. com/ matlabcentral/ fileexchange/64939-cauchy-schwarz-for-vectors$ *and set the vector*

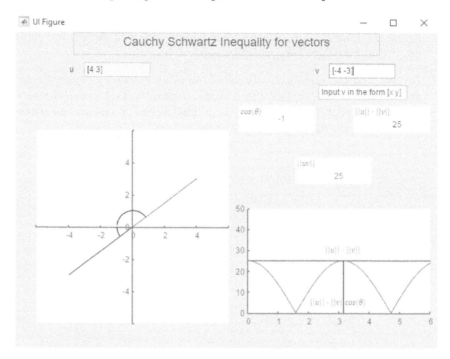

FIGURE 4.3

$v = [x \; y]$ *with a variety of values for* x *and* y *while noting both* $||u|| \cdot ||v||$ *and* $| < u, v > |$. *In general, which quantity did you observe to be larger?*

b. *Determine the angle between* u *and* v *when* $||u|| \cdot ||v|| = | < u, v > |$.

c. *If* V *is the vector space of continuous functions on* $[-1,1]$ *and* $< f,g >= \int_{-1}^{1} f(x)g(x)dx$, *if* $f(x) = x^2$ *and* $g(x) = x$ *does the Cauchy–Schwarz Inequality,* $||f|| \cdot ||g|| \geq | < f,g > |$, *hold?*

d. *Use the demonstration* $https://www.mathworks.com/matlabcentral/$ $fileexchange/64954-cauchy-schwarz-inequality-for-integrals$ *to visualize the Cauchy–Schwarz Inequality for other functions.*

Change of Coordinates of a Vector

Recall that a basis for a vector space is not unique. If B_1 and B_2 are bases for the same vector space V and v_1 is a vector in V written in terms of the vectors in basis B_1, we should also be able to write v_1 in terms of the the basis vectors in B_2. We call this a *change of coordinates* from basis B_1 to basis B_2.

If $B_1 = \{v_1, v_2 \cdots, v_n\}$ and $B_2 = \{u_1, u_2, \cdots, u_n\}$ are two distinct bases

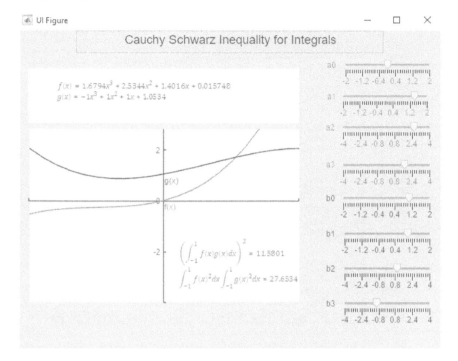

FIGURE 4.4

for R^n, each of the vectors in B_2 can be written as a linear combination of vectors in B_1.

For example, for u_i in B_2, $u_i = k_1v_1 + k_2v_2 + \cdots + k_nv_n$ where $k_1, k_2, \cdots, k_n$ are scalars. And thus you will see in changing from basis B_1 to basis B_2 there are n equations. The matrix of scalar constants resulting from this change of basis is called the *change of coordinates matrix*.

Exercises:

a. *The standard basis for R^2 is $\{(1,0),(0,1)\}$. The set $S_1 = \{(1,2),(-1,2)\}$ is also a basis for R^2. Write each of the basis vectors in S_1 as a linear combination of the standard basis vectors and determine the change of coordinates matrix.*

b. *The vector $\begin{pmatrix} 1 \\ -1 \end{pmatrix} = \begin{pmatrix} 1 & 0 \\ 0 & 1 \end{pmatrix} \begin{pmatrix} 1 \\ -1 \end{pmatrix}$. Rewrite $(1,-1)$ relative to the basis S_1 using the change of coordinates matrix in part a.*

c. *$S_2 = \{(1,2),(-2,1)\}$ is also a basis for R^2. Find the change of coordinates matrix from S_1 to S_2 and use the matrix to write $(1,-1)$ relative to the basis S_2.*

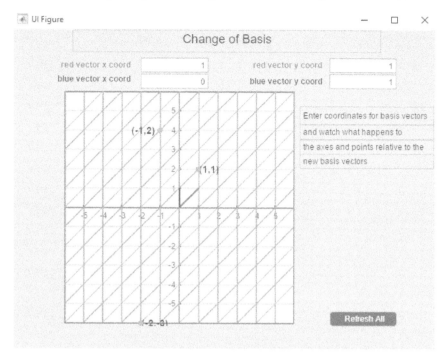

FIGURE 4.5

In the demonstration https://www.mathworks.com/matlabcentral/ fileexchange/64955-change-of-basis the basis vectors, red and blue arrows, are set to (0,1) and (1,0), the standard basis for R^2.

d. *Set the basis vectors to* $(-1,1)$ *and* $(2,2)$. *Are these basis vectors orthogonal?*

e. *When you set the new basis vectors in part d, what happens to the point* $(1,1)$ *relative to the coordinate systems?*

Lab 18: Orthogonal Matrices, QR Decomposition, and Least Squares Regression

Introduction

An *orthogonal matrix* is a square matrix whose column vectors are vectors of magnitude 1 and are pairwise orthogonal (in addition, the row vectors are vectors of magnitude 1 and are pairwise orthogonal).

Exercises:

a. Define $A = \begin{pmatrix} 1 & 0 \\ 0 & 1 \end{pmatrix}$, $B = \begin{pmatrix} \frac{1}{2} & \frac{\sqrt{3}}{2} \\ -\frac{\sqrt{3}}{2} & \frac{1}{2} \end{pmatrix}$, and $M = \begin{pmatrix} 1 & 2 \\ -2 & 1 \end{pmatrix}$.
   Which of these matrices are orthogonal matrices?

b. What is the determinant of each of the orthogonal matrices in part a?

QR Decomposition of Matrices

The QR Decomposition is the decomposition of a matrix $A = QR$ into the product of an orthogonal matrix, Q, and an upper triangular matrix, R. Below we apply the Gram–Schmidt process to create a QR Decomposition for real matrix A.

Assume that the columns, $a_1, a_2, \cdots, a_n$, of A are the vectors $v_1, v_2, v_3, \cdots, v_n$, in the Gram–Schmidt algorithm. Follow the general rule that

$$u_n = v_n - \sum_{i=1}^{n-1} (proj_{u_i} v_n)$$

and $e_n = \frac{u_n}{||u_n||}$, where $e_1, e_2, \cdots, e_n$ will be the columns of the orthogonal matrix Q. The upper triangular matrix R is defined as

$$R = \begin{pmatrix} <e_1,a_1> & <e_1,a_2> & <e_1,a_3> & \ldots & <e_1,a_n> \\ 0 & <e_2,a_2> & <e_2,a_3> & \ldots & <e_2,a_n> \\ 0 & 0 & <e_3,a_3> & \ldots & <e_3,a_n> \\ 0 & 0 & 0 & \ldots & <e_4,a_n> \\ \vdots & \vdots & \vdots & \ddots & \vdots \\ 0 & 0 & 0 & \ldots & <e_n,a_n> \end{pmatrix}.$$

Exercises:

a. Define $A = \begin{pmatrix} 1 & 4 & 8 \\ 2 & 5 & 9 \\ 2 & 6 & 1 \end{pmatrix}$. Use the Gram–Schmidt Process to find Q and R.

b. *Is the QR Decomposition applicable to only square matrices A? To find the QR Decomposition using MATLAB type:*

$$[Q,R]=qr(\textbf{The Name of the Matrix,0})$$

Use this command to find the QR Decomposition of $\begin{pmatrix} 1 & 2 & 3 \\ 4 & 5 & 6 \end{pmatrix}$.

Application to Linear Regression

The goal in Least Squares Linear Regression is to fit a linear function to a data set while minimizing the sum of the residuals squared. To see how the sum of squared errors (or residuals squared) is affected by the choice of the line fitting the given data, use the demonstration https://www.mathworks.com/ matlabcentral/fileexchange/64960-least-square-linear-regression.

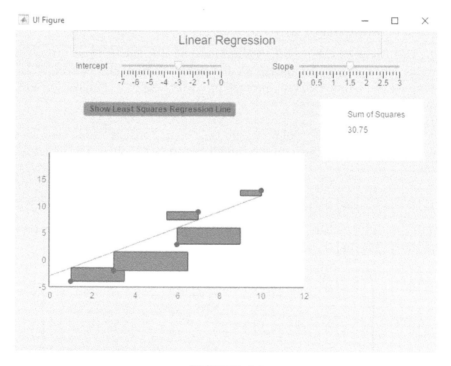

FIGURE 4.6

Exercise: Discuss how the problem of finding a line $y = b + ax$ to fit the given data $\{(x_1,y_1),(x_2,y_2),\cdots,(x_n,y_n)\}$ can be modeled with the overdetermined

$$\text{system} \begin{pmatrix} 1 & x_1 \\ 1 & x_2 \\ 1 & x_3 \\ \vdots & \vdots \\ 1 & x_n \end{pmatrix} \begin{pmatrix} b \\ a \end{pmatrix} = \begin{pmatrix} y_1 \\ y_2 \\ y_3 \\ \vdots \\ y_n \end{pmatrix}. \text{ We will refer to this system in the}$$

form $Ax = b$.

To find the solution (the coefficients for the line), calculate the QR Decomposition of the matrix A, $Ax = QRx = b$, So $x = R^{-1}Q^T b$.

Example: In March 2013, NPR reported that Dunkin' Donuts plans to change its recipes setting a goal of using only 100% sustainable palm oil in making its donuts. The production of palm oil has contributed to large deforestation of rainforests throughout the world as well as issues of erosion and flooding that directly affects the livelihood and lifestyle of the local communities surrounding the rainforest. In addition, the land clearing of Indonesia, which is the largest contributor to the palm oil industry, was 80% illegal in 2008 and has been directly linked to the fate of the wild orangutan.

In this project, we will explore the progression of land clearing for palm oil plantations and its effects on the Sumatran orangutan population.

One can see that there are 7 data points and one of the goals of this project is to fit the best fit line to $y = b_1 x + b_0$ to the data in Table 4.1. Thus, we have an over-determined system with 7 equations and 2 unknowns when we put the points into the linear function. In finding the best fit line, we wish to find the line that minimizes the sum of the squared errors.

TABLE 4.1

Orangutan Population versus Palm Oil Plantations

Year	Total Hectares of Palm Oil Plantations	Sumatran Orangutan Population in Indonesia
2000	4,158,077	13,500
2001	4,713,431	11,245
2002	5,067,058	10,254
2003	5,283,557	8,700
2004	5,566,635	7,500
2005	5,950,349	7,200
2006	6,250,460	6,000

Exercises:

a. *Plot the hectares of palm oil plantations versus the population of the Suma-tran orangutan in Indonesia.*

b. *Use the method described above, with QR Decomposition to determine a line to fit to the data. Plot this line with the data on the same graph.*

Least Squares Regression Take 2

We explored linear regression with QR Decomposition above. Here we will explore other methods for finding a "best fit line." Recall the goal is to minimize the sum of the squared residuals.

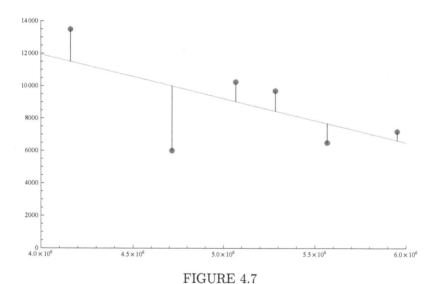

FIGURE 4.7

Define the data as $\vec{b}$ and the estimation line as $A\vec{x}$. Thus the $proj_{A\vec{x}}\vec{b}$ represent the residual error, $\vec{r}$. So $\vec{r}$ is orthogonal to the columnspace of A, $A^T\vec{r} = 0$. Also $\vec{r} = \vec{b} - A\vec{x}$.

Thus $A^T(\vec{b} - A\vec{x}) = 0$. Solving for $\vec{x}$, $A^T\vec{b} = A^T A\vec{x}$ and $\vec{x} = (A^T A)^{-1} A^T\vec{b}$.

Exercises:

a. *Using the data set for the hectares of palm oil plantations versus the Suma-tran orangutan population in Indonesia, from Table 4.1, calculate the "best fit line" using the equation for $\vec{x}$ above.*

b. *Graph the data and the line that you calculated.*

   c. *Compare your results with the "best fit line" using QR Decomposition. Explain why you get the same results.*

Theorems and Problems

For each of these statements, either prove that the statement is true or find a counter example that shows it is false.

Problem 67. If A is an orthogonal square matrix then $A^T = A^{-1}$.

Problem 68. If A is an orthogonal square matrix then $|A| = 1$.

Lab 19: Symmetric Matrices and Quadratic Forms

Introduction

Recall that a square matrix, A, is symmetric if $A = A^T$. Also recall that a square matrix A is diagonalizable if there exists a matrix P such that $D = P^{-1}AP$, where D is a diagonal matrix. In order to diagonalize A, find the eigenvalues of A and the basis for the eigenspace associated with each eigenvalue. The matrix P has columns which are the basis vectors and P diagonalizes A if these vectors are linearly independent.

Just as the basis vectors for an eigenspace are not unique, the matrix P that diagonalizes A is not unique.

Exercises:

a. Let $A = \begin{pmatrix} 3 & 2 & -1 \\ 2 & 3 & -1 \\ -1 & -1 & 4 \end{pmatrix}$. *Determine if A is diagonalizable. If it is diagonalizable, use the eigenvectors to determine a matrix P that diagonalizes A.*

b. *If v_1, v_2, v_3 are the eigenvectors, calculate $v_1 \cdot v_2$, $v_1 \cdot v_3$, and $v_2 \cdot v_3$. What property do the eigenvectors have? Is this property true for all matrices? Is it true for all symmetric matrices?*

c. *Normalize the eigenvectors of A by multiplying by a scalar to make their norm equal to 1. Use these vectors of length one as the columns of a new matrix P. Determine if this new P diagonalizes A. In this case, since A is symmetric, $P^{-1} = P^T$ and thus $P^T AP = D$ and A is orthogonally diagonalizable.*

Quadratic Forms

A *quadratic form* is a function on R^n where $Q_A(x) = x^T A x$, or $Q_A(x_1, x_2, x_3, \cdots, x_n) = \sum_{i \leq j} a_{ij} x_i x_j$, and A is a symmetric matrix. The matrix A is called the *matrix for the form*. Note $Q_A(\vec{0}) = 0$. For example $Q_A : R^2 \to R^2$ defined by $Q_A(x_1, x_2) = x_1^2 + 2x_1 x_2 + x_2^2$ is a quadratic form.

Exercises:

a. *Is $Q_A(x_1, x_2) = x_1^2 + 2x_1 x_2 + x_2^2$ a linear transformation?*

b. *If $x = (x_1, x_2)$, write $Q_A(x_1, x_2) = x_1^2 + 2x_1 x_2 + x_2^2$ in terms of x, x^T and the matrix $A = \begin{pmatrix} 1 & 1 \\ 1 & 1 \end{pmatrix}$.*

Change of Variables in Quadratic Forms

If x is a variable vector in R^n then a change of variables can be represented by $x = Py$, where P is an invertible matrix and y is a new variable vector in R^n. A change of variables in a quadratic form $Q_A(x) = x^T A x$ looks like $Q_A(y) = (Py)^T A(Py) = (y^T)(P^T AP)y$. Since A is symmetric $Q_A(y) = y^T D y$ where D is a diagonal matrix with the eigenvalues of A as the entries on the diagonal.

Exercises:

a. *Determine the quadratic form with matrix for the form*

$$A = \begin{pmatrix} 3 & 2 & -1 \\ 2 & 3 & -1 \\ -1 & -1 & 4 \end{pmatrix}.$$

b. *Diagonalize A using its normalized eigenvectors to create P and determine the quadratic form related to this change of variables.*

c. *Let $A = \begin{pmatrix} 1 & 1 \\ 1 & 1 \end{pmatrix}$ and diagonalize A using its normalized eigenvectors to create P and determine the quadratic form related to this change of variables.*

d. *If $x = R^3$ the cross-product terms are $x_1 x_2$, $x_1 x_3$, $x_2 x_3$ and similarly if $x = R^2$ the cross-product term is $x_1 x_2$. What property related to the cross-product terms do the quadratic forms resulting in the change of variables in parts b. and c. have that the original quadratic forms do not?*

Principal Axes Theorem

Let A be a symmetric matrix. Then there exists a change of variables $x = Py$ that transforms the quadratic form into a quadratic form with no cross-product terms. The columns of P are called the *principal axes* in the change of variables.

Example: Let $Q_A(x,y) = 3x^2 - 4xy + 3y^2$ with the matrix for the form $A = \begin{pmatrix} 3 & -2 \\ -2 & 3 \end{pmatrix}$. The eigenvalues of A are 1 and 5. Let $P = \begin{pmatrix} -\frac{1}{\sqrt{2}} & \frac{1}{\sqrt{2}} \\ \frac{1}{\sqrt{2}} & \frac{1}{\sqrt{2}} \end{pmatrix}$. The new quadratic form is

$$Q_A(x,y) = 5x^2 + y^2.$$

We can geometrically visualize this change of variables. The original quadratic form has a rotated axis where the quadratic form with a diagonal matrix for the form has the standard axes seen in Figure 4.8. To produce the graph in Figure 4.8, type:

```
x = linspace(-4,4);
y = linspace(-4,4);
[X,Y] = meshgrid(x,y);
Z1 = 3*X.^2-4*X.*Y+3*Y.^2;
Z2=5*X.^2+Y.^2;
Z3=Y-X;
Z4=X+Y;
figure
hold on
contour(X,Y,Z1,[10,10],'Color','red')
contour(X,Y,Z2,[10,10],'Color','black')
contour(X,Y,Z3,[0,0])
contour(X,Y,Z4,[0,0])
hold off
```

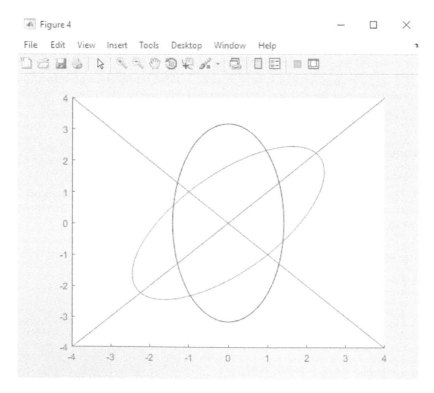

FIGURE 4.8

Properties of Quadratic Forms

A quadratic form is *positive definite* if $Q_A(\vec{x}) > 0$ for all $\vec{x} \neq \vec{0}$ and *negative definite* if $Q_A(\vec{x}) < 0$ for all $\vec{x} \neq \vec{0}$. If $Q_A(\vec{x})$ takes on both positive and negative values then it is called *indefinite*. A quadratic form is called *semipositive definite* if it never takes on negative values. Similarly it is called *seminegative definite* if it never takes on positive values.

Exercises:

a. *Give an example of a positive definite quadratic form on* R^2.

b. *Give an example of a negative definite quadratic form on* R^3.

c. *Find the eigenvalues of the matrix for the form, A, in your examples in parts a. and b. What do you conjecture about the sign of the eigenvalues of A?*

A real symmetric $n \times n$ matrix, A, is called
1) positive definite if $x^T A x > 0$ for all x in R^n,
2) semipositive definite if $x^T A x \geq 0$ for all x in R^n,
3) negative definite if $x^T A x < 0$ for all x in R^n and
4) seminegative definite if $x^T A x \leq 0$ for all x in R^n.

Exercises:

a. *To graph* $Q_A(x_1,x_2) = x_1^2 + x_2^2$ *when* $Q_A(x_1,x_2) = 1$ *type*
   x = linspace(-1,1);
   y = linspace(-1,1);
   [X, Y] = meshgrid(x,y);
   Z = X.^2+Y.^2;
   contour(X,Y,Z,[1,1])
   Is the matrix for the form, A, positive or negative definite? How does this relate to the shape of the graph?

b. *Use the demonstration* $https://www.mathworks.com/matlabcentral/fileexchange/64976-conic-sections$ *to look at the graph of* $Q_A(x_1,x_2) = ax_1^2 + bx_2^2$ *when* $Q_A(x_1,x_2) = 1$ *where a and b are constants. For each behavior difference that you discover write down the matrix for the form A and calculate the eigenvalues for each A.*

c. *What eigenvalues determine an ellipse and which determine a hyperbola?*

d. *Determine whether the curve* $2x^2 + 10xy - y^2 = 1$ *is an ellipse or hyperbola. What is the matrix for the form (possibly not symmetric) affiliated with this conic section when you complete the square?*

e. *Use the* **plot3** *command to plot* $Q_A(x,y) = 2x^2 + 10xy - y^2$. *Does this function have a saddle point, global maximum, or global minimum? Where*

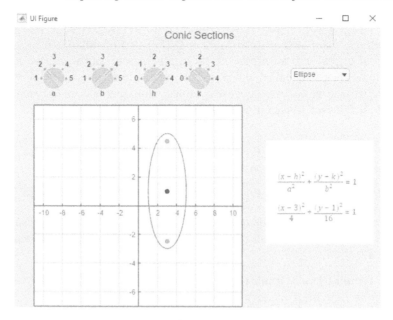

FIGURE 4.9

is this point located? (Note: If $Q_A(x) = x^T A x$ and A is invertible then $x = (0,0)$ is the only critical point and thus is the saddle point, global maximum or global minimum.)

Theorems and Problems

For each of these statements, either prove that the statement is true or find a counter example that shows it is false.

Theorem 69. If A is symmetric then any two eigenvectors of A are orthogonal.

Theorem 70. If A is symmetric then A is orthogonally diagonalizable.

Problem 71. If $Q(x_1, x_2, x_3, \cdots, x_n)$ is a quadratic form with all real coefficients then it is positive definite if and only if

$$Q(x_1, x_2, x_3, \cdots, x_n) = x_1^2 + x_2^2 + x_3^2 + \cdots + x_n^2.$$

Theorem 72. A quadratic form Q is positive definite if and only if the eigenvalues of the coefficient matrix A are all positive.

Theorem 73. If A is a positive or negative definite matrix then A is invertible.

Problem 74. If A is a symmetric 2×2 matrix with eigenvalues $\lambda_1 \geq \lambda_2$ and Q_A is the quadratic form defined by $Q_A(x) = x^T A x$, then the conic section defined by $Q_A(x) = 1$ is

(1) an ellipse if $\lambda_1 \geq \lambda_2 > 0$,

(2) a hyperbola if $\lambda_1 > 0 > \lambda_2$,

(3) the empty set if $0 \geq \lambda_1 \geq \lambda_2$ and

(4) two parallel lines if $\lambda_1 > \lambda_2 = 0$.

Project Set 4

Project 1: Lights Out

The 5×5 Lights Out game was explored in Project Set 1 and 2 where you created the adjacency matrix, initial state vector, final state vector, and solutions. In Project Set 2, your final exploration was to look at the 5×5 Lights Out game where the buttons can take on three states, 0, 1, and 2, and the goal of the game is to go from an initial state of all lights in state 0 and end with a final state of all lights in state 1.

1	2	3	4	5
6	7	8	9	10
11	12	13	14	15
16	17	18	19	20
21	22	23	24	25

a. Look back at your solution to Project Set 2, Project 1, part d. Which games have invertible adjacency matrices that quickly lead to solutions? If the adjacency matrix was not invertible, did the game have a solution? If not, how did you determine that no solution exists?

b. If the $5 \times n$ game has adjacency matrix, M, that is not invertible modulo 3, you may have determined that there was no solution. We will explore this further. Choose an n such that the adjacency matrix, M, for the $5 \times n$ game is not invertible modulo 3 and determine a basis for the nullspace of M. To explore the nullspace, nullpress, vectors for this matrix further you may wish to visit https://www.mathworks.com/matlabcentral/fileexchange/65109-multistate-lights-out.

c. Recall that if there is a solution, the goal is to determine a push vector p such that $Mp + i = f$ where $i = \vec{0}$ and $f = \vec{1}$. Another way to think about this problem is that we wish to determine p such that $\vec{1}$ is in the range of the transformation defined by multiplying by the standard matrix M. Restate this statement in terms of the columnspace of M.

d. Note that matrix M is symmetric. In part c, you stated the problem in terms of the columnspace of M; now restate the problem in terms of the rowspace of M.

e. Restate the statement in part d in terms of the nullspace of M.

f. For the $5 \times n$ game that you chose in part b, calculate the Euclidean inner product of each nullspace vector, nullpress vectors, with $\vec{1}$ modulo 3. Based on these results, does the game have a solution?

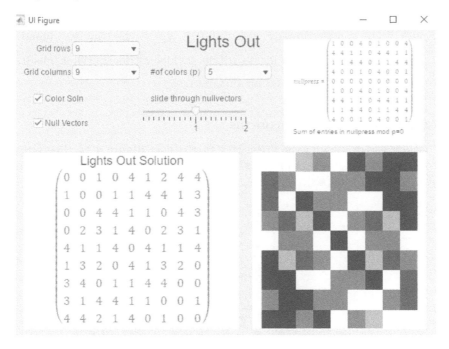

FIGURE 4.10

g. Continue to explore which $5 \times n$ games with 3 colors have solutions based on this new knowledge and write up your results.

Project 2: Linear Regression

In Lab 18, you learned several ways to determine a "best fit" line for the data related to hectares of oil palm plantations and the population of Sumatran orangutans in Indonesia.

a. Use the same ideas to fit a quadratic function to the data in Lab 18. (Hint: Think about what the matrix A should be.)

b. Use the same ideas to fit a cubic function to the data in Lab 18.

c. One way to determine the best estimation is to calculate the error of your estimates. There are several errors that you can calculate. Let $\hat{y}(x_i)$ be the approximate y value given by plugging the x value x_i into the model\function, and b_i is the exact y value corresponding to the value x_i given by the data.

The maximum error, denoted $||\hat{y} - b||_\infty$ is the maximum difference in magnitude between the exact data and the approximation, $\max_{1 \le i \le n} |\hat{y}_i - b_i|$, where n is the number of data points. The l_2 error denoted $||\hat{y} - b||_2 = \sqrt{\sum_{i=1}^{n}(\hat{y}_i - b_i)^2}$. Finally the relative l_2 error can be found by $\frac{\sqrt{\sum_{i=1}^{n}(\hat{y}_i - b_i)^2}}{\sqrt{\sum_{i=1}^{n}b_i^2}}$. Use these ideas to discuss which estimation (line, quadratic, or cubic) is the best model for this data.

Project 3: Cosine Transforms

Consider the graphic of the green tree frog call in Figure 4.11. The frog call looks periodic like $\cos(x)$ or $\sin(x)$ but there is not any one function of the form $\cos(kx)$ or $\sin(kx)$, with $k \in R$ that can describe the call.

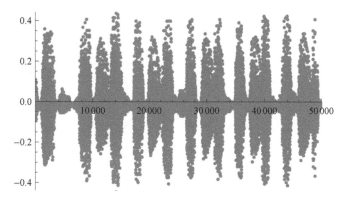

FIGURE 4.11: Green tree frog sound wave

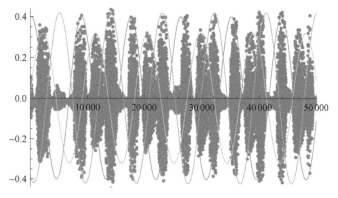

FIGURE 4.12: Cosine graphs over sound wave

a. By inspection of the period of the repetitions in the call, determine k_1, k_2 and k_3, where $\cos(k_1x)$, $\cos(k_2x)$ and $\cos(k_3x)$ are shown in Figure 4.12.

b. Graph $\cos(k_1 x)$, $\cos(k_2 x)$, $\cos(k_3 x)$ and $\cos(k_1 x) + \cos(k_2 x) + \cos(k_3 x)$, from part a. Which function is the best approximation to the frog call?

c. A set of functions $f_1(x), f_2(x), \ldots, f_n(x)$ are linearly independent if and only if the

$$\text{Wronskian} = \begin{vmatrix} f_1(x) & f_2(x) & \cdots & f_n(x) \\ f_1'(x) & f_2'(x) & \cdots & f_n'(x) \\ \vdots & \vdots & \ddots & \vdots \\ f_1^{(n-1)}(x) & f_2^{(n-1)}(x) & \cdots & f_n^{(n-1)}(x) \end{vmatrix}$$

is not equal to 0 for all $x \in R$. Determine if the functions $\cos(k_1 x)$, $\cos(k_2 x)$ and $\cos(k_3 x)$ from part a. are linearly independent.

d. We can write a continuous function $f(x)$ as an infinite series of cosine functions called the *Fourier Cosine Series* on the interval $-L \le x \le L$, $f(x) = \sum_{n=0}^{\infty} A_n \cos\left(\frac{n\pi x}{L}\right)$. The coefficients A_n are constant real numbers. The set of continuous functions is a vector space. Based on the Fourier Cosine Series, describe a basis for the vector space of continuous functions on the interval $-L \le x \le L$.

e. Define the inner product on functions which are continuous from -1 to 1 as

$$< f,g >= \int_{-1}^{1} f(x)g(x)dx.$$

Is the basis $\{1, \cos(\pi x), \cos(2\pi x), \cdots, \cos(k\pi x), \cdots\}$ for any integer k an orthogonal basis? Is it an orthonormal basis?

The green tree frog call is recorded as a discrete set of data so we cannot write it as a Fourier Cosine Series, but when we collect data with noise we can use the Discrete Fourier Cosine Transform to try to get rid of the noise.

Project 4: The Hadamard Product on Matrices

For two matrices A and B of the same size, the Hadamard product $A \circ B$ is defined as $(A \circ B)_{ij} = a_{ij} \cdot b_{ij}$.

Define $A = \begin{pmatrix} 1 & 2 \\ 2 & 4 \end{pmatrix}$, $B = \begin{pmatrix} 1 & 3 \\ 3 & 9 \end{pmatrix}$, and $M = \begin{pmatrix} -1 & 1 \\ 2 & -2 \end{pmatrix}$.

a. Calculate $A \circ B$. Note that **A.*B** in MATLAB calculates the Hadamard product between A and B.

b. Calculate $A \circ (B + M)$ and $A \circ B + A \circ M$. Are they equal?

c. Is the Hadamard product commutative, $A \circ B = B \circ A$?

d. If A_1 is an upper triangular matrix and A_2 is any matrix of the same size as A_1, determine what type of matrix results in $A_1 \circ A_2$.

e. Is $M_{2,2}$ under the Hadamard product, representing the defined matrix addition, and traditional scalar multiplication a vector space?

f. Which of the matrices A, B, and/or M are semipositive definite? (Recall that an $n \times n$ matrix, A, is semipositive definite if $x^T A x \geq 0$ for all $x \in R^n$.) Find the Hadamard product of those matrices which are semipositive definite and determine if the resulting matrix has any special qualities.

g. The pth *Hadamard power* of a matrix A has $(i,j)^{th}$ entry equal to a_{ij}^p. Is the pth Hadamard power of a semipositive definite matrix semipositive definite?

h. Do similar results to those found in parts f. and g. hold for seminegative definite matrices?

i. Summarize your findings about the Hadamard product.

Project 5: Hadamard Matrices and Image Compression

A *Hadamard matrix*, H, is an $n \times n$ matrix whose entries are either -1 or 1 such that $HH^T = nI$.

a. Give an example of a 2×2 Hadamard matrix, H_1, that is invertible.

b. Give an different example of a Hadamard matrix, H_2, such that $tr(H_2) = 0$.

c. Write a short MATLAB code to generate all 2×2 Hadamard matrices and then visualize them by altering H in the following code.
```
H=[-1 1;1 -1];
map=[1 1 1;0 0 0];
colormap(map);
image(2*H)
```

d. Using the Euclidean inner product for R^2, find the inner product between the columns, and rows, of each of the 2×2 Hadamard matrices. What property do these Hadamard matrices have?

e. Change your program slightly to find and visualize 3×3 Hadamard matrices. Keep in mind it is possible that for some value of n, there are no $n \times n$ Hadamard matrices.

f. If H_1 from part a. is a 2×2 Hadamard matrix, are the 4×4 matrix

$$\begin{pmatrix} H_1 & H_1 \\ -H_1 & H_1 \end{pmatrix} \text{ and the } 8 \times 8 \text{ matrix } \begin{pmatrix} H_1 & H_1 & H_1 & H_1 \\ -H_1 & H_1 & -H_1 & H_1 \\ -H_1 & -H_1 & H_1 & H_1 \\ H_1 & -H_1 & -H_1 & H_1 \end{pmatrix}$$

Hadamard matrices? If so, plot H_1, the 4×4, and 8×8 matrix to see a pattern.

Image compression is the process of taking a high-quality image and, for the sake of transfer or storage, reducing the size of the image, getting rid of any redundancies. In order to do this, one must first determine what part of the image is most important to the image quality. One way to determine this is through the use of Hadamard matrices, or *Hadamard transformations*.

We show a 1-D example here, using the Hadamard transformation matrix $A = \frac{1}{\sqrt{2}} \begin{pmatrix} 1 & 1 \\ -1 & 1 \end{pmatrix}$.

g. If the original image vector is $\vec{v} = \begin{pmatrix} v_1 \\ v_2 \end{pmatrix} = \begin{pmatrix} 4 \\ 6 \end{pmatrix}$, use the transformation matrix A to transform $\vec{v}$, and determine which component v_1 or v_2 is more significant based on their transformed size.

In 2-D, instead of using columns of 1's and -1's in A we use images created by Hadamard matrices. For example, a 2×2 image would be transformed using transformation matrix

$$\begin{pmatrix} 1 & 1 & -1 & 1 \\ 1 & 1 & -1 & 1 \\ -1 & -1 & 1 & -1 \\ 1 & 1 & -1 & 1 \end{pmatrix} =$$

5

Matrix Decomposition with Applications

Lab 20: Singular Value Decomposition (SVD)

Introduction

Although rarely tackled at the undergraduate level, SVD is extremely useful, particularly in statistics and signal processing. In Lab 19, we looked at square matrices that are diagonalizable and orthogonally diagonalizable. An important fact about the diagonalization is the resulting diagonal matrix contains the eigenvalues of the original matrix on the main diagonal. However, not all matrices are diagonalizable. In this case you may look at singular value decomposition (SVD). If A is an $m \times n$ matrix, singular values, σ_j, are the square roots of the eigenvalues for the matrix $A^T A$.

The term singular value relates to the distance of the given matrix to a singular matrix. The idea behind SVD is that every matrix A, can be decomposed into the product $U \sum V^T$ where U and V are orthogonal matrices and $\sum_{ii} = \sigma_i$ and $\sum_{ij} = 0$ otherwise.

Recall from Lab 19 that all symmetric matrices are orthogonally diagonalizable. Thus for all matrices A, $A^T A$ is symmetric, we can find an orthogonal matrix P such that $A^T A = PDP^T$.

SVD is extremely unique in that it can be found for all matrices and it can be used to find the best/optimal k-rank approximation of a matrix.

Calculating the SVD

The SVD for an $m \times n$ matrix A, $A = U \sum V^T$ where U is a $m \times m$ orthogonal matrix whose columns form an orthonormal basis for R^m, V is an $n \times n$ orthogonal matrix whose columns form an orthonormal basis for R^n, and $\sum$ is an $m \times n$ matrix such that $\sum_{ii} = \sigma_i$.

Since all symmetric matrices are orthogonally diagonalizable, we can find an orthogonal matrix P such that

$$A^T A = PDP^T = V \sum U^T U \sum V^T = V \begin{pmatrix} \sigma_1^2 & 0 & 0 & 0 \\ 0 & \sigma_2^2 & 0 & 0 \\ 0 & 0 & \ddots & 0 \\ 0 & 0 & 0 & \sigma_n^2 \end{pmatrix} V^T$$

and

$$AA^T = U \sum V^T V \sum U^T = U \begin{pmatrix} \sigma_1^2 & 0 & 0 & 0 \\ 0 & \sigma_2^2 & 0 & 0 \\ 0 & 0 & \ddots & 0 \\ 0 & 0 & 0 & \sigma_m^2 \end{pmatrix} U^T.$$

Note also that $A^T A v_i = \sigma_i^2 v_i$, $AA^T u_i = \sigma_i^2 u_i$, and $A v_i = \sigma_i u_i$.

Example: Find the SVD of $A = \begin{pmatrix} -1 & 0 & 3 \\ 0 & 2 & 0 \end{pmatrix}$.

$A^T A = \begin{pmatrix} 1 & 0 & -3 \\ 0 & 4 & 0 \\ -3 & 0 & 9 \end{pmatrix}$ with eigenvalues 10, 4, and 0.

Using normalized eigenvectors of $A^T A$, define $V = \begin{pmatrix} \frac{-1}{\sqrt{10}} & 0 & \frac{3}{\sqrt{10}} \\ 0 & 1 & 0 \\ \frac{3}{\sqrt{10}} & 0 & \frac{1}{\sqrt{10}} \end{pmatrix}$. Sim-

ilarly use the normalized eigenvectors of $AA^T = \begin{pmatrix} 10 & 0 \\ 0 & 4 \end{pmatrix}$ to define

$U = \begin{pmatrix} 1 & 0 \\ 0 & 1 \end{pmatrix}$. Finally define $\sum = \begin{pmatrix} \sqrt{10} & 0 & 0 \\ 0 & 2 & 0 \end{pmatrix}$.

Exercise: Find the SVD for $\begin{pmatrix} 1 & 1 \\ 0 & 0 \end{pmatrix}$.

Orthogonal Grids: Visualizing SVD

Here we look at a visualization of singular values. We begin by visualizing the transformation with square matrices. Just as we have learned how to apply linear transformations to vectors in R^2, we can explore what happens if we apply those same transformations to the Cartesian grid. If the transformed grid lines remain orthogonal, we call this transformed grid an *orthogonal grid*. The following demonstration shows how linear transformations affect the orthogonality of grid lines.

Exercises:

a. *To determine if rotation or dilation alone change the orthogonality of the grid, use* https://www.mathworks.com/matlabcentral/fileexchange/65197-orthogonal-grids.

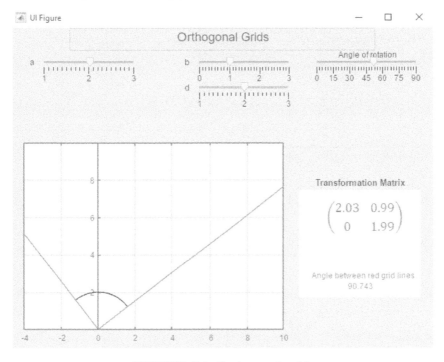

FIGURE 5.1: Orthogonal grids

b. *Setting $a = 2, b = 1$, and $d = 2$, determine the approximate angle of rotation, θ, that produces an orthogonal grid with axes defined by the red vectors in the demonstration. This particular transformation is called a sheer transformation.*

For the following exercises use the demonstration $https://www.mathworks.com/matlabcentral/fileexchange/65264-singular-values$.

c. *Denote the original (blue) vectors, in the demonstration, as v_1 and v_2, using the same sheer transformation described in b., determine the approximate lengths of the transformed (red) vectors, Mv_1 and Mv_2, when the sheer grid axes are orthogonal. These lengths are called the singular values, σ_1 and σ_2, of M.*

d. *Vectors u_1 and u_2 are orthonormal vectors in the direction of Mv_1 and Mv_2 respectively when Mv_1 and Mv_2 are orthogonal. Find u_1 and u_2.*

Orthogonal Components of a Vector

The orthogonal components of a vector $v = v_w + v_{w^\perp}$ where v_w in W and $v_{w^\perp}$ in $W^\perp$, the orthogonal component to W. Given an orthonormal basis

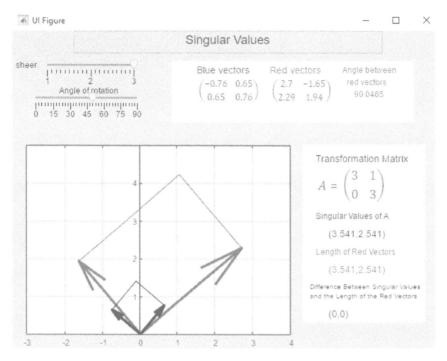

FIGURE 5.2: Singular values related to sheer transformation

$\{v_1, v_2, \cdots, v_n\}$ for W, $v_w = (v_1 \cdot v)v_1 + (v_2 \cdot v)v_2 + \cdots + (v_n \cdot v)v_n$.

Using the theory above, for any vector x, $x = (v_1 \cdot x)v_1 + (v_2 \cdot x)v_2$ and thus $Mx = M(v_1 \cdot x)v_1 + M(v_2 \cdot x)v_2 = u_1\sigma_1(v_1 \cdot x) + u_2\sigma_2(v_2 \cdot x)$.

Noting that for any two vectors u and w, $u \cdot w = u^T w$, we can say that $Mx = u_1\sigma_1(v_1^T x) + u_2\sigma_2(v_2^T x)$.

More generally $M = U \sum V^T$ where U is a matrix whose columns are the vectors u_1 and u_2, $\sum_{ii} = \sigma_i$ and $\sum_{ij} = 0$ otherwise, and V is a matrix whose columns are the vectors v_1 and v_2. This is called the *singular value decomposition* of M.

Exercise: Using your results from c and d above, determine U, $\sum$, and V such that $U \sum V^T = M$ where M is the shear matrix $\begin{pmatrix} k & 1 \\ 0 & k \end{pmatrix}$, $k = 2$.

Relating Eigenvalues and Singular Values

Recall that an eigenvector, x, is the solution to $(A - \lambda I)x = 0$, where λ is an eigenvalue and A is a square matrix. This system of homogeneous equations has a solution precisely when $A - \lambda I$ is singular. We have gone into detail about eigenvalues and the corresponding eigenvectors of square matrices in Lab 14, but is there a similar concept for matrices which are not square?

In general, eigenvalues and singular values are not related except when the matrix is symmetric. If a matrix A is symmetric then its singular values are the absolute values of its eigenvalues.

Note also that if A is symmetric that the eigenvectors of A are the same as the eigenvectors of $A^T A$ and AA^T and thus the normalized eigenvectors of A can be used to define V and U.

Exercises: Define $A = \begin{pmatrix} 25 & 15 \\ 15 & 25 \end{pmatrix}$.

a. *Determine the eigenvalues and singular values of A. Use the singular values to define $\sum$.*

b. *Find the eigenvectors of A and determine V and U.*

Application to Data Imaging: Reducing Noise

SVD is regularly used to smooth out noisy data in such problems as imaging. Essentially by not including all of the singular values in the singular value decomposition, one can begin to eliminate the noise in a data set.

Exercises:

a. *Define the data set,* **data = [0 1 0 1 0 1 0 1 0 1;1 0 1 0 1 0 1 0 1 0;0 1 0 1 0 1 0 1 0 1;1 0 1 0 1 0 1 0 1 0; 0 1 0 1 0 1 0 1 0 1;1 0 1 0 1 0 1 0 1 0;0 1 0 1 0 1 0 1 0 1;1 0 1 0 1 0 1 0 1 0;0 1 0 1 0 1 0 1 0 1;1 0 1 0 1 0 1 0 1 0]**; *and Type* **colormap(gray);image(75*data)** *to see the data set without noise.*

b. *Define a noisy data set and use the image command to visualize the noisy data. Type* **r = -.4 + (.4+.4)*rand(10,10); noisy=r+data;image(75*noisy);** *This noisy data set will be your matrix M for the SVD algorithm. The noisy data set is the original data with some random noise added in.*

c. *Define M = noisy, and Type* **svd(M)** *to see a list of all of the singular values for M. Determine the dominant singular values (and even more* *importantly the number of dominant singular values). These will be the ones that you will include in your SVD to reduce the noise in the data.*

d. *Type $[U, W, V] = svds(M,n)$, where n is the number of dominant singular values you wish to include in the SVD (determined in part c.). Note here the matrices for the SVD will be stored in U, W, and V. Multiply $U * W * transpose(V)$ to find an improved data set with reduced noise. Use the image command to visualize this improved data. (You may want to type $image(75*U*W*transpose(V))$ in order to see the contrast.)*

SVD is also applied extensively to the study of linear inverse problems and is useful in the analysis of regularization methods such as that of Tikhonov. It is widely used in statistics where it is related to principal component analysis and in signal processing and pattern recognition.

Lab 21: Cholesky Decomposition and Its Application to Statistics

LU Decomposition and Doolittle Decomposition

The LU Decomposition algorithm is used to write a square matrix, A, as a product of a lower and an upper triangular matrix. The applications of both the LU and Cholesky Decomposition will be discussed later on in the lab.

Recall from Lab 2, that we can perform elementary row operations on a matrix by multiplying the matrix by elementary matrices. The steps of the LU Decomposition of matrix A are to

1. Determine elementary matrices to transform A into an upper triangular matrix, U. That is find $E_1, E_2, \cdots, E_k$ such that $E_k \cdots E_2 E_1 A = U$.

2. Write $A = (E_k \cdots E_2 E_1)^{-1} U = LU$ where $L = (E_k \cdots E_2 E_1)^{-1}$.

If A is also symmetric, then $LU = A = A^T = (LU)^T = U^T L^T$. Thus $U(L^T)^{-1} = L^{-1} U^T = D$, and $U = DL^T$, where D is a diagonal matrix. Thus if A is symmetric we can find L and D such that $A = LDL^T$. This is called the *Doolittle Decomposition* of A.

If A is Hermitian (Lab 9), then $LU = A = \overline{A}^T = \overline{LU}^T = \overline{U}^T \overline{L}^T$. Therefore $U(\overline{L}^T)^{-1} = \overline{L}^{-1} \overline{U}^T = D$ and we can write $A = LD\overline{L}^T$.

Exercises:

a. Define $A = \begin{pmatrix} 2 & 1 \\ 1 & 2 \end{pmatrix}$. *Find the LU Decomposition and the Doolittle Decomposition of A using the algorithm above.*

b. Define $B = \begin{pmatrix} 2 & 1+2i \\ 1-2i & 4 \end{pmatrix}$. *Find the LU Decomposition and the Doolittle Decomposition of B.*

The MATLAB command:

[L,U,P] = lu(The Name of the Matrix)

produces L a lower triangular matrix, U, and upper triangular matrix, and P, a permutation matrix, such that $LU = PA$, where A is the original matrix.

If all of the entries in D are positive (and thus A is positive definite) then we can find the Cholesky Decomposition of A.

The Cholesky Decomposition Algorithm

If A is symmetric with real entries or Hermitian and if A is positive definite (Lab 19) then there exists a lower triangular matrix with nonnegative diagonal entries such that $A = L\overline{L}$. L is considered to be the square root of A. This decomposition of the matrix A is called the *Cholesky Decomposition*.

It is helpful to note that a matrix is positive definite if and only if all of its eigenvalues are positive.

The Cholesky Decomposition Algorithm:

1. Write $A = L\overline{L}^T$ as $\begin{pmatrix} a_{11} & \overline{A_{21}}^T \\ A_{21} & A_{22} \end{pmatrix} = \begin{pmatrix} l_{11} & 0 \\ L_{21} & L_{22} \end{pmatrix} \begin{pmatrix} l_{11} & \overline{L_{21}}^T \\ 0 & \overline{L_{22}}^T \end{pmatrix}$ and $l_{11} = \sqrt{a_{11}}$.

2. In general for real matrices $l_{ii} = \sqrt{a_{ii} - \sum_{j=1}^{i-1} l_{ij}^2}$ and
   $l_{ik} = \frac{1}{l_{kk}}(a_{ik} - \sum_{j=1}^{k-1} l_{ij}l_{kj})$ for $i > k$.
   For Hermitian Matrices $l_{ii} = \sqrt{a_{ii} - \sum_{j=1}^{i-1} l_{ij}\overline{l_{ij}}}$ and
   $l_{ik} = \frac{1}{l_{kk}}(a_{ik} - \sum_{j=1}^{k-1} l_{ij}\overline{l_{kj}})$ for $i > k$.

The Cholesky Decomposition is more efficient than the LU Decomposition Algorithm for symmetric matrices and is a modified form of Gaussian Elimination.

Exercises: Let $A = \begin{pmatrix} 2 & 1 \\ 1 & 2 \end{pmatrix}$ *and* $B = \begin{pmatrix} 2 & 1+2i \\ 1-2i & 4 \end{pmatrix}$.

a. *Determine if A is positive definite.*

b. *Use the algorithm above to find the Cholesky Decomposition of A.*

c. *Determine if B is positive definite. If B is positive definite find the Cholesky Decomposition of B.*

d. *What is the relationship between the Cholesky Decomposition and the Doolittle Decomposition?*

To use MATLAB to produce a lower triangular matrix for the Cholesky Decomposition, type:

L = chol(The Name of the Matrix,'lower').

Generating Random Correlated Data Using Cholesky Decomposition

The goal in this section is to see one way that decomposition of matrices applies to statistics, particularly related to the generation of correlated data.

The *covariance matrix* captures the variance and linear correlation in multivariable data. *Covariance* is a measure of how much m data sets (of size N) change together. The variance, σ_i, in the ith data set is shown on the main diagonal. The covariance matrix,

$$
\text{Cov} = \begin{pmatrix}
\sum \frac{x_1^2}{N} & \sum \frac{x_1 x_2}{N} & \sum \frac{x_1 x_3}{N} & \cdots & \sum \frac{x_1 x_m}{N} \\
\sum \frac{x_2 x_1}{N} & \sum \frac{x_2^2}{N} & \sum \frac{x_2 x_3}{N} & \cdots & \sum \frac{x_2 x_m}{N} \\
\sum \frac{x_3 x_1}{N} & \sum \frac{x_3 x_2}{N} & \sum \frac{x_3^2}{N} & \cdots & \sum \frac{x_3 x_m}{N} \\
\vdots & \vdots & \vdots & \ddots & \vdots \\
\sum \frac{x_m x_1}{N} & \sum \frac{x_m x_2}{N} & \sum \frac{x_m x_3}{N} & \cdots & \sum \frac{x_m^2}{N}
\end{pmatrix}.
$$

Notice that the covariance matrix is a symmetric matrix.

Many times when you are analyzing data, it is common to have to work with lots of variables. Sometimes these variables are redundant (or related) and there are only a few true sources (or variables of relevance) of information in the data. It is an analyst's job to determine those sources.

The *correlation matrix* is directly related to the covariance matrix, as it lists the correlation coefficients between two random variables i and j in the ij^{th} entry; however the main diagonal entries will have a value of 1 representing a full positive linear correlation between a variable and itself.

Exercises:

a. *The data that is in Table 5.1 is International Monetary Fund (IMF) data for 6 world regions related to GDP and distribution of GDP. Find the covariance matrix related to the given variables. Type var(data set name) to find the variance of a data set and to find the covariance matrix affiliated with n data sets type cov(data set 1; data set 2;...;data set n).*

b. *The entries of the correlation matrix, Cor, are directly related to the covariance matrix. $Cor(i,j) = \frac{cov(i,j)}{\sigma_i \sigma_j}$. Find the correlation matrix for the data above.*

The correlation between two variables x and y, or correlation coefficient, will always be between -1 and 1. If the correlation coefficient is close to 1 then

TABLE 5.1

IMF Data

	GDP Current Prices	Investment % of GDP	Gross National Savings % of GDP	Volume of Exports of Goods % of GDP
Central and Eastern Europe	1844.682	21.222	17.005	4.802
Commonwealth of Independent States	2658.841	24.362	27.591	4.081
Developing Asia	12324.727	41.891	42.972	4.424
ASEAN 5	1935.796	29.827	30.624	2.822
Latin America and the Caribbean	5765.563	21.442	19.549	3.456
Middle East, North Africa, Afghanistan, and Pakistan	3422.987	24.964	36.157	3.885

we say that there is a positive linear correlation, that is, when one variable increases the other variable increases.

If the correlation coefficient is close to -1 then we say that there is a negative linear correlation, and thus when one of the variables increases the other will decrease. If the correlation coefficient is close to 0 then there is no linear correlation.

c. *Using your correlation matrix from part b. find the correlation coefficient between the Investment % of the GDP and the Gross national savings % of the GDP and interpret the results.*

d. *Find the correlation coefficient between Investment % of GDP and Volume of exports of goods and interpret this result.*

e. *Find the Cholesky Decomposition of the correlation matrix from part b.*

In the next few exercises, we will use the Cholesky Decomposition of the correlation matrix to find a list of correlated data. Note that the correlation of the data is also dependent on the strength of correlation among the variables in your matrix. Therefore, a set of variables which are significantly correlated would produce even more dramatic results than those from this example. We will start with a random set of data and then use the decomposition to correlate it.

f. *Generate a random table of 1200 real data points, example Type:*

$$r = -1 + (1+1)*rand(6,200);$$

and plot the data by typing
$x = linspace(1,800,800);$
$scatter(x,[r(1,:) \ r(2,:) \ r(3,:) \ r(4,:)]).$

g. Using Matrix L, from part e., calculate $C = L * r$ to get your correlated data. To see a graph of this data Type:

$$scatter(x,[C(1,:) \ C(2,:) \ C(3,:) \ C(4,:)])$$

If the correlated data $C = L.r$ is actually more linearly correlated, you will see a less random and more linear behavior in this second plot.

Lab 22: Jordan Canonical Form

Introduction

A *Jordan Block* is of the form
$$\begin{pmatrix} \lambda & 1 & 0 & 0 & 0 & 0 \\ 0 & \lambda & 1 & 0 & 0 & 0 \\ 0 & 0 & \lambda & 1 & 0 & 0 \\ \vdots & \vdots & \vdots & \ddots & \ddots & \vdots \\ 0 & 0 & 0 & 0 & \lambda & 1 \\ 0 & 0 & 0 & 0 & 0 & \lambda \end{pmatrix}.$$

A square matrix A is in *Jordan canonical form*, or *Jordan normal form*, if there are Jordan block matrices J_i, $i = 1, 2, \cdots, k$ such that

$$A = \begin{pmatrix} J_1 & 0 & 0 & 0 & 0 \\ 0 & J_2 & 0 & 0 & 0 \\ 0 & 0 & J_3 & 0 & 0 \\ \vdots & \vdots & \vdots & \ddots & \vdots \\ 0 & 0 & 0 & 0 & J_k \end{pmatrix}.$$

Similar to diagonalizing a matrix, we say that a matrix, A, is in a *Jordan canonical form*, J, if there is a matrix P such that $J = P^{-1}AP$.

Generalized Eigenvectors

If a matrix A is not diagonalizable it may be because the algebraic multiplicity of one or more eigenvalues is greater than 1 and the eigenvectors are linearly dependent. We can however find the Jordan canonical form of A using generalized eigenvectors.

If matrix A has eigenvalue λ with algebraic multiplicity k, then vector v is a *generalized eigenvector* of rank k associated with λ if and only if $(A - \lambda\mathrm{I})^k v = 0$ and $(A - \lambda\mathrm{I})^{k-1} v \neq 0$. If matrix A has eigenvalue λ with algebraic multiplicity k, then there are k linearly independent generalized eigenvectors associated with λ. Note that a generalized eigenvector of rank 1 is the same as an eigenvector.

Example: Let $A = \begin{pmatrix} 2 & 3 & 1 \\ 0 & 2 & -1 \\ 0 & 0 & 2 \end{pmatrix}$. A has eigenvalue $\lambda = 2$ with algebraic multiplicity 3. Compute the original eigenvector, $v_1 = (1,0,0)$.

To find the generalized eigenvector of rank 2, we use the original eigenvector v_1 to solve $(A - 2I)^2 v_2 = 0$. Note that we wish to solve $(A - 2I)^2 v_2 = (A - 2I)v_1 = 0$ or $(A - 2I)v_2 = v_1$ and thus $v_2 = (0, \frac{1}{3}, 0)$.

Similarly, to find the generalized eigenvector of rank 3, solve $(A - 2I)^3 v_3 = 0$ or $(A - 2I)v_3 = v_2$, $v_3 = (0, \frac{1}{9}, -\frac{1}{3})$.

Thus $P = \begin{pmatrix} 1 & 0 & 0 \\ 0 & \frac{1}{3} & \frac{1}{9} \\ 0 & 0 & -\frac{1}{3} \end{pmatrix}$ and $P^{-1}AP = \begin{pmatrix} 2 & 1 & 0 \\ 0 & 2 & 1 \\ 0 & 0 & 2 \end{pmatrix}$ which is the Jordan canonical form of A. We call v_1, v_2, and v_3 the *chain of generalized eigenvectors* associated with $\lambda = 2$.

In order to find the Jordan canonical form using MATLAB type:

[V,J]=jordan(The Name of the Matrix);

Here the matrix J is the Jordan canonical form of the original matrix, V holds the generalized eigenvectors, and $V^{-1}AV = J$.

Exercises:

a. Let $A = \begin{pmatrix} 1 & 2 & 3 & 4 \\ 0 & 1 & 7 & 8 \\ 0 & 0 & 6 & 12 \\ 0 & 0 & 0 & 6 \end{pmatrix}$, *is A diagonalizable?*

b. *The matrix A, from part a, may not be diagonalizable but may be "almost" diagonalizable in its Jordan canonical form. Determine the generalized eigenvectors and the Jordan canonical form of A.*

c. *Determine how the values in the Jordan canonical blocks relate to eigenvalues of A.*

d. *Find the Jordan canonical form of A^2 and relate it to the Jordan canonical form of A.*

e. *Find the Jordan canonical form of e^A and relate it to the Jordan canonical form of A.*

f. *Let A be a 5×5 matrix with two distinct eigenvalues λ of multiplicity 3 and μ of multiplicity 2. Determine all possible Jordan canonical forms of A up to permutations of the Jordan blocks.*

The Minimal Polynomial

Cayley–Hamilton's Theorem tells us that if a square matrix A has characteristic equation $p(x) = 0$ then $p(A) = 0$. The *minimal polynomial* affiliated with matrix A, $p_m(x)$ is a unique monic, strictly increasing or strictly decreasing, polynomial of least degree such that $p_m(A) = 0$.

If $\lambda_1, \lambda_2, \ldots, \lambda_k$ are distinct eigenvalues of A with s_i as the largest Jordan block affiliated with λ_i then the minimal polynomial is degree $\sum s_i$.

Example: For $A = \begin{pmatrix} 1 & 0 & 0 & 0 \\ 0 & 1 & 0 & 0 \\ 1 & -1 & 1 & 0 \\ 1 & -1 & 1 & 1 \end{pmatrix}$, A has characteristic polynomial

$p(x) = (x - 1)^4$ and Jordan canonical form $\begin{pmatrix} 1 & 1 & 0 & 0 \\ 0 & 1 & 1 & 0 \\ 0 & 0 & 1 & 0 \\ 0 & 0 & 0 & 1 \end{pmatrix}$. Since the

largest Jordan block affiliated with $\lambda = 1$ is a 3×3 block, the minimal polynomial is degree 3. In this case notice that $p(A) = (A - I)^4 = 0$ and $p_m(A) = (A - I)^3 = 0$.

Exercises: Let $A = \begin{pmatrix} 1 & 2 & 3 & 4 \\ 0 & 1 & 7 & 8 \\ 0 & 0 & 6 & 12 \\ 0 & 0 & 0 & 6 \end{pmatrix}$ and $B = \begin{pmatrix} 4 & 0 & 0 & 0 \\ 2 & 2 & 3 & 0 \\ -1 & 0 & 2 & 0 \\ 1 & 0 & 1 & 2 \end{pmatrix}$.

a. Find the characteristic polynomial of A and determine the degree of the minimal polynomial affiliated with A.

b. Find the characteristic polynomial and minimal polynomial for B.

Theorems and Problems

For each of these statements, either prove that the statement is true or find a counter example that shows it is false.

Problem 75: If the Jordan canonical form of A is $\begin{pmatrix} J_1 & 0 & 0 & 0 & 0 \\ 0 & J_2 & 0 & 0 & 0 \\ 0 & 0 & J_3 & 0 & 0 \\ \vdots & \vdots & \vdots & \ddots & \vdots \\ 0 & 0 & 0 & 0 & J_k \end{pmatrix}$

then the Jordan canonical form of A^m is $\begin{pmatrix} J_1^m & 0 & 0 & 0 & 0 \\ 0 & J_2^m & 0 & 0 & 0 \\ 0 & 0 & J_3^m & 0 & 0 \\ \vdots & \vdots & \vdots & \ddots & \vdots \\ 0 & 0 & 0 & 0 & J_k^m \end{pmatrix}$.

Problem 76: If the Jordan canonical form of A is $\begin{pmatrix} J_1 & 0 & 0 & 0 & 0 \\ 0 & J_2 & 0 & 0 & 0 \\ 0 & 0 & J_3 & 0 & 0 \\ \vdots & \vdots & \vdots & \ddots & \vdots \\ 0 & 0 & 0 & 0 & J_k \end{pmatrix}$

then the Jordan canonical form of e^A is
$$\begin{pmatrix} e^{J_1} & 0 & 0 & 0 & 0 \\ 0 & e^{J_2} & 0 & 0 & 0 \\ 0 & 0 & e^{J_3} & 0 & 0 \\ \vdots & \vdots & \vdots & \ddots & \vdots \\ 0 & 0 & 0 & 0 & e^{J_k} \end{pmatrix}.$$

Theorem 77: If A is invertible then the Jordan canonical blocks of A^{-1} are the same as those for A.

Theorem 78: The chain of generalized eigenvectors, $v_1, v_2, \ldots, v_k$ associated with λ are linearly independent.

Project Set 5

Project 1: Singular Value Decomposition in Text Analysis

We will be analyzing the following quotes for clusters of similarities using SVD.

Quote 1: Education is the most powerful weapon which you can use to change the world. (Mandela)
Quote 2: Education is not the filling of a pail, but the lighting of a fire. (Yeats)
Quote 3: Education is not preparation for life; education is life itself. (Dewey)
Quote 4: Real knowledge is to know the extent of one's ignorance. (Confucius)
Quote 5: Intelligence is the ability to adapt to change. (Hawkings)
Quote 6: It always seems impossible until it's done. (Mandela)
Quote 7: You must be the change you wish to see in the world. (Gandhi)
Quote 8: Information is not knowledge. (Einstein)

a. Create a matrix, A, with quotes as the rows and the possible words (all words without repeats in the 8 quotes) representing the columns. $A_{i,j} =$ Frequency of word j in quote i. For example if "the" is the word represented in column 3 then $A_{1,3} = 2$.

b. Calculate the Singular Value Decomposition of A using only the first two singular values. Plot the points $(U(i,1),U(i,2))$ using scatter and look for quotes that are close together/clustered.

c. This is a small data set. In order to see how to apply SVD to clustering, discuss other places where these methods may be applicable.

Project 2: The Collatz Problem

Let $f(k) = \begin{cases} \frac{3k+1}{2}, & \text{when k is odd,} \\ \frac{k}{2}, & \text{when k is even.} \end{cases}$

The Collatz conjecture is that for each natural number k, the sequence

$$k, f(k), (f \circ f)(k), (f \circ f \circ f)(k), \cdots$$

contains the number 1.

For each natural number n, the $n \times n$ *Collatz matrix* A_n has entries
$$a_{ij} = \begin{cases} 1 & \text{if } i = f(j), \\ 0 & \text{otherwise.} \end{cases}$$

Define the graph G_n as the directed graph with adjacency matrix A_n. A *chain* in G_n is an ordered list of distinct vertices $S = \{v_1, v_2, \cdots, v_m\}$ such that $f(v_i) = v_{i+1}$, for $1 \le i < m$ and $f(v_m) \ne v_1$.

A *cycle* in G_n is a chain where $f(v_m) = v_1$. The length of the chain or cycle is equal to m.

a. Find A_2, A_3, A_4.

b. Explore A_n for larger values of n and make a conjecture about the characteristic equation of A_n.

c. For $n \geq 2$, make a conjecture about the number of Jordan blocks for the eigenvalue 0 of the matrix A_n.

d. For $n \geq 2$, make a conjecture about the maximum length of a cycle in G_n.

Project 3: Generalized Inverses

Throughout this course we have only discussed the inverse of a square matrix. (Recall that a square invertible matrix, A, satisfies $AA^{-1} = I$ and $A^{-1}A = I$.) If A is an $m \times n$ matrix, where m and n are distinct, A is not invertible. However it may be possible to find a matrix B such that $ABA = A$. We call this matrix, B, the *pseudoinverse*, or *generalized inverse*, of A. The generalized inverse of A is denoted A^+. Note that the generalized inverse is not unique.

If $m \geq n$ then the inverse of $A^T A$ exists and $A^+ = (A^T A)^{-1}A^T$. (Here $A^+ A = I$ holds.)

If $m \leq n$ then the inverse of AA^T exists and $A^+ = A^T(AA^T)^{-1}$. (Here $AA^+ = I$ holds.)

Although we only discussed LU decomposition in terms of square matrices, the LU decomposition can be found for non-square matrices.

Given an $m \times n$ matrix A with rank r, there exists a factorization (which we call the LU decomposition) $A = P_r L U P_c$ where P_r is an $m \times m$ permutation matrix, L is an $m \times r$ lower triangular matrix, U is an $r \times n$ upper triangular matrix, and P_c is an $n \times n$ permutation matrix.

The structure of $L = \begin{pmatrix} \mathcal{L}_1 \\ \mathcal{L}_2 \end{pmatrix}$ where $\mathcal{L}_1$ is an $r \times r$ lower triangular invertible matrix. The structure of U is $(\mathcal{U}_1, \mathcal{U}_2)$ where $\mathcal{U}_1$ is an $r \times r$ upper triangular invertible matrix. $A^+ = P_c^T \begin{pmatrix} \mathcal{U}_1^{-1}\mathcal{L}_1^{-1} & 0 \\ 0 & 0 \end{pmatrix} P_r^T$.

a. Use your knowledge of the LU decomposition of square matrices to find the LU decomposition of $A = \begin{pmatrix} 2 & 1 & 2 & 1 \\ 2 & 1 & 3 & 2 \\ 4 & 2 & 6 & 8 \end{pmatrix}$.

b. Use your LU decomposition from part a to find A^+.

c. What might a Cholesky decomposition look like for a non-square matrix? Make a conjecture about Cholesky decomposition for non-square matrices and use your ideas to find this decomposition of A.

d. Discuss how to find a generalized inverse of a non-square matrix using your Cholesky decomposition from part c.

e. Type **pinv(The Name of the Matrix)** to find a generalized inverse for A. When the matrix is not square this command uses SVD to calculate the inverse.

f. We have seen many instances where finding the inverse of a square matrix was important. Give a few examples of where it may be important to find the generalized inverse of a matrix.

g. Compare your results in parts b, d, and e and write a summary of your findings.

Project 4: Singular Value Decomposition and Music Genomics

In Project Set 3, we discussed seriation/ordering, of musical pieces based on similarities and the Fiedler vector. The study of how musical pieces and artists influence each other is called *music genomics*. In this project, we apply singular value decomposition to our study of music genomics. The problem will use the data from Project Set 3, Project 6, Figure 3.6.

Use the binary data that you created in Project Set 3, Project 6 part a or create the data set based on the description given in part a of that project.

a. Create the similarity matrix, S, described in part b. of Project Set 3, Project 6 and compute the singular value decomposition of S keeping only the 2 largest singular values. Type:

$$[\mathbf{U}, \mathbf{W}, \mathbf{V}] = \mathbf{svds(S,2)}.$$

b. With 20 songs in the data set, the matrix U will be a 20×2 matrix. The rows of U are the new coordinates for each of the songs, s_i, based on this decomposition. Plot the new coordinates of the songs to see which songs are closest together based on the SVD algorithm. Note that you might want to create a labeling of the songs before plotting them. To do this type:
labels = cellstr(num2str([1 : length(U)]'));
plot(U(: ,1), U(: ,2),'rx');
text(U(: ,1), U(: ,2), labels)

c. Let us now assume that there is one song s_1 that we wish to compare with all of the others. We will use the cosine distance from the new coordinates of s_1, Michael Jackson's Beat It, to rank the original songs. The distance from s_1 to s_i is $d_i = \frac{s_1 \cdot s_i}{||s_1|| \cdot ||s_i||}$. Find d_i for each song, other than s_1, and rank the songs based on how close they are to s_1.

d. It is possible that an artist has been influenced by Michael Jackson but not his song Beat It. Define q_1 as the average of the new coordinates for s_1 and s_{11}, both Michael Jackson songs. Find the distance from each song to q_1, $d_i = \frac{q_1 \cdot s_i}{||q_1|| \cdot ||s_i||}$ and rank the songs based on how close they are to q_1.

6

Applications to Differential Equations

Lab 23: Linear Differential Equations

System of Linear Differential Equations

What is a differential equation?

A *differential equation* is an equation that relates a function with its derivatives of different orders.

A system of first-order *linear differential equations* is of the form

$$
\begin{aligned}
x_1' &= a_{11}(t)x_1 + a_{12}(t)x_2 + \cdots + a_{1n}(t)x_n + f_1(t) \\
x_2' &= a_{21}(t)x_1 + a_{22}(t)x_2 + \cdots + a_{2n}(t)x_n + f_2(t) \\
x_3' &= a_{31}(t)x_1 + a_{32}(t)x_2 + \cdots + a_{3n}(t)x_n + f_3(t) \\
&\ \ \vdots \\
x_n' &= a_{n1}(t)x_1 + a_{n2}(t)x_2 + \cdots + a_{nn}(t)x_n + f_n(t),
\end{aligned}
$$

where $x_1(t), x_2(t), \ldots, x_n(t)$ are functions of some parameter t. This system of differential equations can be written in the *fundamental form*

$$
\begin{pmatrix} x_1 \\ x_2 \\ x_3 \\ \vdots \\ x_n \end{pmatrix}' =
\begin{pmatrix}
a_{11}(t) & a_{12}(t) & \ldots & a_{1n}(t) \\
a_{21}(t) & a_{22}(t) & \ldots & a_{2n}(t) \\
a_{31}(t) & a_{32}(t) & \ldots & a_{3n}(t) \\
\vdots & \vdots & \ddots & \vdots \\
a_{m1}(t) & a_{m2}(t) & \ldots & a_{mn}(t)
\end{pmatrix}
\begin{pmatrix} x_1 \\ x_2 \\ x_3 \\ \vdots \\ x_n \end{pmatrix} +
\begin{pmatrix} f_1(t) \\ f_2(t) \\ f_3(t) \\ \vdots \\ f_n(t) \end{pmatrix} .
$$

A differential equation may have an associated initial condition, when $t = 0$ as well. The *general solution* of a differential equation, or system of differential equations, is not affiliated with an initial condition and is a family of solutions, where a *particular solution* is affiliated with a particular initial condition. If the system of differential equations is of the form $x' = Ax$ then we say that the differential equation is a *homogeneous system*.

Basic Examples

In order to solve the differential equation $x' = \lambda x$, think of a function whose derivative is a scalar λ times the original function. The general solution is $x(t) = Ce^{\lambda t}$, if the initial condition is $x(0) = 1$, the particular solution is $x(t) = e^{\lambda t}$.

Exercise: Find the solution to $x' = Ax$ where $A = \begin{pmatrix} 1 & 0 \\ 0 & .5 \end{pmatrix}$, $x = \begin{pmatrix} x_1 \\ x_2 \end{pmatrix}$.
Determine the particular solution, $x_1(t)$ and $x_2(t)$, with initial conditions $x_1(0) = .5$ and $x_2(0) = .5$. To plot the solution Type:
t=0:0.1:1;
x1= Solution Function for x_1;
x2= Solution Function for x_2;
plot(x1,x2)

We call this system *uncoupled* because the differential equations related to x_1 and x_2 are not dependent on one another. We will see some *coupled systems* below.

Visualizing General Solutions

Recall the First Derivative test (from Calculus 1).

Given a function $f(x)$
1) if $\frac{d}{dx}f(x) > 0$ then $f(x)$ is increasing,
2) if $\frac{d}{dx}f(x) < 0$ then $f(x)$ is decreasing and
3) if $\frac{d}{dx}f(x) = 0$ then x is a critical value of $f(x)$.

We use this information to draw a vector field, called a *phase portrait*, showing general solutions to a differential equation.

Exercises:

a. *Use the demo www.mathworks.com/matlabcentral/fileexchange/ 64494-homogeneous-systems-of-coupled-linear-differential -equations to visualize the solution to the system $x' = Ax$, where $A = \begin{pmatrix} 1 & 0 \\ 0 & .5 \end{pmatrix}$. You will have to type the initial conditions in for x_0 and y_0 to see the particular solution. You can also click the check box to see many solution curves.*

b. *Determine how the eigenvalues, affiliated with matrix A, are incorporated into the solution to the system of differential equations.*

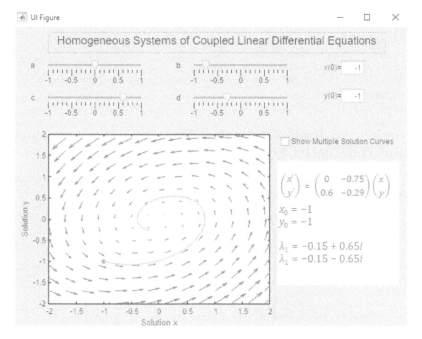

FIGURE 6.1

c. Given that diagonal matrix $A = \begin{pmatrix} a & 0 \\ 0 & d \end{pmatrix}$, determine the general solution to $x' = Ax$.

d. Now let's explore some coupled systems. For $x' = Ax$ when $A = \begin{pmatrix} 0 & -1 \\ .5 & .25 \end{pmatrix}$ the general solution looks like $x = Ce^{\lambda t}$, where λ is the eigenvalues of A. Find λ and determine the general solution.

e. Euler's formula says that $e^{ix} = \cos(x) + i\sin(x)$. Use Euler's formula to rewrite the particular solution to part d when $x_1(0) = .5$ and $x_2(0) = .5$.

f. Part of your solution from part e should look like

$$\begin{aligned} x_1(t) &= C_1 e^{\lambda_1 t} \\ &= C_1 e^{real(\lambda_1)t}(\cos(imag(\lambda_1)t) + i\sin(imag(\lambda_1)t)) \end{aligned}$$

where $real(\lambda_1)$ represents the real part of λ_1 and $imag(\lambda_1)$ represents the imaginary part of λ_1. To see a visual representation of this solution Type:

t=0:.1:10;
x1=C_1 * exp(real(λ_1) * t). * (cos(imag(λ_1)t);

x2=$C_1 * exp(real(\lambda_1) * t). * (sin(imag(\lambda_1)t)$;
plot(x1,x2)

g. *Determine values for a, b, c, and d that would produce circular solution curves.*

Applying the Jordan Canonical Form to a Homogeneous System of Differential Equations

Let $x' = Jx$ be a differential equation where the $k \times k$ Jordan block J is associated with eigenvalue λ. Then the solution to $x' = Jx$ is

$$x(t) = e^{\lambda t} \begin{pmatrix} 1 & t & \frac{t^2}{2!} & \frac{t^3}{3!} & \cdots & \frac{t^{k-1}}{(k-1)!} \\ 0 & 1 & t & \frac{t^2}{2!} & \cdots & \frac{t^{k-2}}{(k-2)!} \\ 0 & 0 & 1 & t & \cdots & \frac{t^{k-3}}{(k-3)!} \\ \vdots & \vdots & \ddots & \ddots & \ddots & \vdots \\ 0 & 0 & \cdots & 0 & 1 & t \\ 0 & 0 & 0 & \cdots & 0 & 1 \end{pmatrix} C$$

where $C = \begin{pmatrix} c_1 \\ c_2 \\ \vdots \\ c_k \end{pmatrix}$ is a vector of arbitrary constants.

If the system of differential equations is $x' = Ax$ where A is not in Jordan canonical form, we can use the Jordan canonical form of A, $J = P^{-1}AP$, to solve the system.

The idea behind this is that $x' = Ax$, $x' = PJP^{-1}x$ and thus $(P^{-1}x)' = J(P^{-1}x)$. If $y = P^{-1}x$ the system becomes $y' = Jy$ and the techniques from the previous section can be used to find y. Then $x = Py$.

Exercises:

a. *Let* $J = \begin{pmatrix} 2 & 1 & 0 \\ 0 & 2 & 1 \\ 0 & 0 & 2 \end{pmatrix}$ *be a Jordan block associated with* $\lambda = 2$. *Solve* $x' = Jx$ *where* $x(0) = \begin{pmatrix} 1 \\ 0 \\ 0 \end{pmatrix}$.

b. Solve the system $x' = Ax$, where $A = \begin{pmatrix} 2 & 3 & 1 \\ 0 & 2 & -1 \\ 0 & 0 & 2 \end{pmatrix}$. Use the Jordan canonical form of A to find the solution to the system of differential equations.

Lab 24: Higher-Order Linear Differential Equations

Introduction

A general n^{th} order linear differential linear equation is of the form $G(t) = P_n(t)x^{(n)} + P_{(n-1)}(t)x^{(n-1)} + \cdots + P_2(t)x'' + P_1(t)x' + P_0(t)x$.

Let's look at converting a certain type of higher-order linear differential equation into a system of first-order equations. We will be looking specifically at higher-order linear homogeneous differential equations with $P_n(t) = 1$.

Let $x_1 = x$, $x_2 = x' = x_1'$, $\cdots$, $x_n = x^{(n-1)} = x_{n-1}'$, then the nth order homogeneous equation becomes the system

$$
\begin{pmatrix} x_1 \\ x_2 \\ \vdots \\ x_{n-1} \\ x_n \end{pmatrix}' = \begin{pmatrix} 0 & 1 & 0 & \cdots & 0 \\ 0 & 0 & 1 & \cdots & \vdots \\ \vdots & \vdots & \vdots & \vdots & 1 \\ -P_0(t) & -P_1(t) & -P_2(t) & \cdots & -P_{n-1}(t) \end{pmatrix} \begin{pmatrix} x_1 \\ x_2 \\ \vdots \\ x_{n-1} \\ x_n \end{pmatrix}.
$$

We will designate this system as $x' = Ax$. Recall that $Ce^{\lambda t}$ is a solution to $x' = Ax$. If an eigenvalue, λ, of A has multiplicity k, then $C_1 e^{\lambda t}$, $C_2 t e^{\lambda t}$, $C_3 t^2 e^{\lambda t}$, $\cdots$, $C_{k-1} t^{k-2} e^{\lambda t}$ and $C_k t^{k-1} e^{\lambda t}$ are solutions to $x' = Ax$.

In addition if $x_1(t)$ and $x_2(t)$ are solutions to $x' = Ax$ then so is $(x_1+x_2)(t)$.

Exercises:

a. *Rewrite the homogeneous third-order linear differential equation $x^{(3)} - 4x'' + 4x' = 0$ as a system of first-order equations.*

b. *Find the eigenvalues and characteristic equation of A.*

c. *Use the eigenvalues to find a general solution to the differential equation from part a.*

d. *Rewrite the fourth-order linear differential equation*

$$x^{(4)} + 8x'' + 16x = 0$$

as a system of first-order equations. Then use the eigenvalues to find a general solution to the differential equation.

Applying Cramer's Rule to Solve Nonhomogeneous Systems

Cramer's Rule

If A is a square $n \times n$ invertible matrix, the solution to $Ax = b$ is

$$x_1 = \frac{|A_1|}{|A|}, x_2 = \frac{|A_2|}{|A|}, \cdots, x_n = \frac{|A_n|}{|A|}$$

where A_i is the $n \times n$ matrix created by replacing the ith column of A with the vector b.

Solving Nonhomogeneous n^{th}-Order Linear Order Systems Using Cramer's Rule

The fundamental form for the differential equations

$$x^{(3)} - 2x'' - 21x' - 18x = 0$$

is
$\begin{pmatrix} x_1 \\ x_2 \\ x_3 \end{pmatrix}' = \begin{pmatrix} 0 & 1 & 0 \\ 0 & 0 & 1 \\ 18 & 21 & 2 \end{pmatrix} \begin{pmatrix} x_1 \\ x_2 \\ x_3 \end{pmatrix}$, denoted $x' = Ax$.

Matrix A has the characteristic equation $(-6 + r)(1 + r)(3 + r) = 0$ and thus the general solution is $x(t) = C_1 e^{-3t} + C_2 e^{-t} + C_3 e^{6t}$.

If the ith component of the general solution to the homogeneous solution is denoted y_i, then the solution to the nonhomogeneous system of differential $x^{(3)} - 2x'' - 21x' - 18x = t$ is of the form $X_p(t) = u_1 y_1 + u_2 y_2 + u_3 y_3$.

It is important to note that, in this process, we assume that $u_1' y_1 + u_2' y_2 + u_3' y_3 = 0$ and $u_1' y_1' + u_2' y_2' + u_3' y_3' = 0$.

Thus $X_p'(t) = u_1 y_1' + u_2 y_2' + u_3 y_3'$ and $X_p''(t) = u_1 y_1'' + u_2 y_2'' + u_3 y_3''$. Plugging all of these equations back into the original differential equations we get
$u_1(y_1^{(3)} - 2y_1'' - 21y_1' - 18y_1) + u_2(y_2^{(3)} - 2y_2'' - 21y_2' - 18y_2) + u_3(y_3^{(3)} - 2y_3'' - 21y_3' - 18y_3) + u_1'(y_1'') + u_2' y_2'' + u_3' y_3'' = t$.

Since y_1, y_2, and y_3 are all solutions to the original homogeneous equation, the above equation becomes

$$u_1' y_1'' + u_2' y_2'' + u_3' y_3'' = t.$$

The goal becomes to find u_1, u_2 and u_3 satisfying $u_1' y_1 + u_2' y_2 + u_3' y_3 = 0$, $u_1' y_1' + u_2' y_2' + u_3' y_3' = 0$, and $u_1' y_1'' + u_2' y_2'' + u_3' y_3'' = t$.

That is, to find $X_p(t)$ the goal is to find u_1, u_2, and u_3 such that

$$\begin{pmatrix} y_1 & y_2 & y_3 \\ y_1' & y_2' & y_3' \\ y_1'' & y_2'' & y_3'' \end{pmatrix} \begin{pmatrix} u_1 \\ u_2 \\ u_3 \end{pmatrix}' = \begin{pmatrix} 0 \\ 0 \\ t \end{pmatrix}. \text{ This is of the form } Ax' = b.$$

Exercises:

a. *For the system $x^{(3)} - 2x'' - 21x' - 18x = t$, determine y_1, y_2 and y_3.*

b. *Determine A in the system $Ax' = b$ affiliated with the differential equation in part a and use Cramer's Rule to solve for u_1', u_2', and u_3'.*

c. *Use your solution for u_1, u_2, and u_3 in part b and integrate each component to find u_1, u_2, and u_3.*

d. *Use your results from a to find the particular solution to the nonhomogeneous system $X_p(t) = u_1 y_1 + u_2 y_2 + u_3 y_3$.*

e. *Find the general solution to the nonhomogeneous system which is $x(t) + X_p(t)$, where $x(t)$ is the solution to $x^{(3)} - 2x'' - 21x' - 18x = 0$.*

Theorems and Problems

For each of these statements, either prove that the statement is true or find a counter example that shows it is false.

Problem 79. A linear combination of solutions to the differential equation $x' = Ax$ is also a solution.

Problem 80. If an eigenvalue, λ, of A has multiplicity k, then $C_1 e^{\lambda t}$, $C_2 t e^{\lambda t}$, $C_3 t^2 e^{\lambda t}, \cdots, C_{k-1} t^{k-2} e^{\lambda t}$ and $C_k t^{k-1} e^{\lambda t}$ are solutions to $x' = Ax$.

Lab 25: Phase Portraits, Using the Jacobian Matrix to Look Closer at Equilibria

Given the system of differential equations

$$
\begin{aligned}
x_1' &= f_1(x_1, x_2, x_3, \ldots, x_n) \\
x_2' &= f_2(x_1, x_2, x_3, \ldots, x_n) \\
x_3' &= f_3(x_1, x_2, x_3, \ldots, x_n) \\
&\vdots \\
x_n' &= f_n(x_1, x_2, x_3, \ldots, x_n).
\end{aligned}
$$

The *Jacobian matrix*, A, has entries $A_{ij} = \frac{\partial}{\partial x_j} f_i$.

If $f_1(x_1, x_2) = (x_1 + 1)\sin(x_2)$, we can find $\frac{\partial}{\partial x_2} f_1$ using MATLAB, by typing
syms x1 x2
f1=(x1+1)*sin(x2);
diff(f1,x2).

Nullclines and Equilibrium points

The *nullclines* of a system are the curves determined by solving $f_i = 0$ for any i. The *equilibrium points* of the system, or the fixed points of the system, are the point(s) where the nullclines intersect.

The equilibrium point is said to be *hyperbolic* if all eigenvalues of the Jacobian matrix have non-zero real parts. In a two-dimensional system, a hyperbolic equilibrium is called a *node* when both eigenvalues are real and of the same sign. If both of the eigenvalues are negative then the node is stable, or a *sink*, and unstable when they are both positive, or a *source*.

A hyperbolic equilibrium is called a *saddle* when eigenvalues are real and of opposite signs.

When eigenvalues are complex conjugates then the equilibrium point is called a *spiral point*, or *focus*. This equilibrium point is stable when the eigenvalues have a real part which is negative and unstable when they have positive real part.

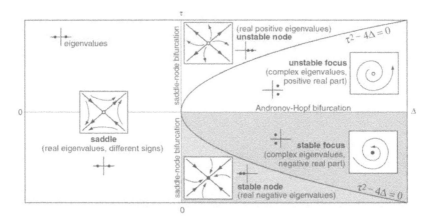

FIGURE 6.2: *http : //www.scholarpedia.org/article/Equilibrium*

Exercises:

a. *Determine the Jacobian matrix associated with the system*

$$x' = -x + y, y' = -6x + \frac{1}{2}y.$$

b. *Find the equilibrium points of the system and the eigenvalues of the Jacobian matrix in part a. and use Figure 6.2 to determine the type(s) of equilibrium points that are present in the system.*

c. *Use the demo www.mathworks.com/matlabcentral/fileexchange/ 64580-visualizing-the-solutions-of-two-linear-differential -equations to visualize how the equilibrium point(s) from part b behave. You can check on the Phase Portrait checkbox in the demonstration as well to see a different look to the system. Describe the behavior that you see.*

d. *Determine the Jacobian matrix associated with the system*

$$x' = -\frac{1}{2}x + 3y, y' = -6x + y.$$

Find the equilibrium points of the system and the eigenvalues of the Jacobian matrix to determine the type of equilibrium points that are present.

e. *Use the demonstration from part c to visualize how the equilibrium point(s) behave. Describe the behavior that you see.*

f. *Determine the Jacobian matrix associated with the nonlinear system*

$$x' = x(4 - 2x - y), \ y' = y(5 - x - y).$$

g. *Determine the nullclines and equilibrium points of the system in part f.*

h. *Find the Jacobian matrix of the system, in part f, at each of the equilibrium points. Then find the eigenvalues of each of these Jacobian matrices to determine what type of equilibrium points are present in the system.*

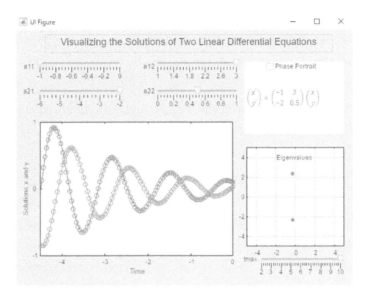

FIGURE 6.3

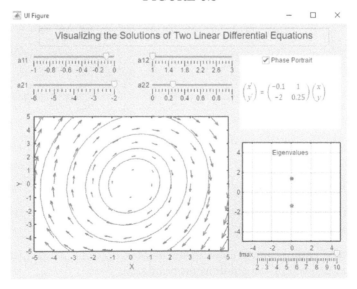

FIGURE 6.4

Project Set 6

Project 1: Predator Prey Model

This system of nonlinear differential equations models the populations of two species in a closed system: one species is the predator (ex. shark) and one is the prey (ex. fish). If $x(t)$ denotes the prey population and $y(t)$ the predator population, the differential model is of the form:

$$\frac{dx}{dt} = x(a - by), \quad \frac{dy}{dt} = -y(c - dx),$$

where a and c are growth parameters and b and d are interaction parameters.

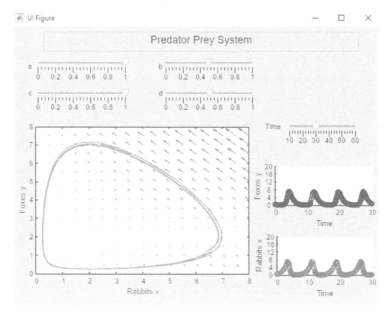

FIGURE 6.5: Visualizing the predator prey behavior

a. Determine what happens to the system in the absence of prey and in the absence of the predator.

b. Find the equilibrium points (in terms of $a, b, c,$ and d) and the Jacobian matrix at each of the equilibrium points.

c. Determine the behaviors of the solutions at each of the equilibrium points.

d. Choose a set of parameters (values for $a, b, c,$ and d) and write a synopsis of the solution curves related to these parameters. Use the demonstration at `www.mathworks.com/matlabcentral/fileexchange/` `64676-predator-prey-system` to help you visualize what is happening with your parameters.

Project 2: Lorenz Equations Applied to Finance

The Lorenz system of nonlinear differential equations,

$$\frac{dx}{dt} = \sigma(y - x), \frac{dy}{dt} = x(\rho - z), \frac{dz}{dt} = xy - \beta z,$$

sometimes represents chaotic behavior in different disciplines.

The nonlinear chaotic financial system can be described similarly with the system

(Equation 1) $\dfrac{dx}{dt} = \left(\dfrac{1}{b} - a\right)x + z + xy,$

(Equation 2) $\dfrac{dy}{dt} = -by - x^2,$

(Equation 3) $\dfrac{dz}{dt} = -x - cz,$

where x represents interest rate in the model, y represents the investment demand, and z is the price exponent. In addition, the parameter a represents savings, b represents per-investment cost, and c represents elasticity of demands of commercials.

We will explore this system in two different parts.

a. Looking only at Equations 1 and 2, find the equilibrium point(s) when $ab \geq 1$ and use the Jacobian matrix to determine what type of equilibrium point(s) are present.

b. Looking only at Equations 1 and 2, find the equilibrium point(s) when $ab < 1$ and use the Jacobian matrix to determine what type of equilibrium point(s) are present.

c. Looking only at Equations 2 and 3, find the equilibrium point(s) when $x = 0$ and use the Jacobian matrix to determine what type of equilibrium point(s) are present.

d. Looking only at Equations 2 and 3, find the equilibrium point(s) when $x \neq 0$ and use the Jacobian matrix to determine what type of equilibrium point(s) are present.

e. Set the parameters $a = 0.00001, b = 0.1$, and $c = 1$. Graph the solution by finding the numerical solution to the system, Type:

$a = 0.00001;$

$b = 0.1;$

$c = 1;$

$F = @(t,y)[(1/b-a)*y(1)+y(3)+y(1)*y(2); -b*y(2)-y(1)*y(1); -y(1)-c*y(3)];$

$[t,y] = ode45(F,[0,180],[.1,.2,.3]);$

$plot3(y(:,1),y(:,2),y(:,3));$

$xlabel('Interest Rate(x)');$

$ylabel('Investment Demand(y)');$

$zlabel('Price Exponent(z)');$

$axis\ tight;$

f. Write an analysis of the graph of the solution based on your analysis in parts a-d. Note that you can rotate the 3D graph by clicking on the tool circled in Figure 6.6, clicking on the graph and moving the mouse simultaneously. If you wish to see the graph as it moves through time

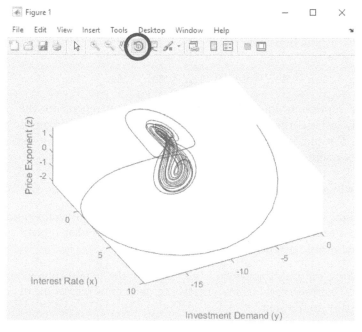

FIGURE 6.6: Visualizing the Lorenz Equations Using the Rotate 3D Tool

type:

$a = 0.00001;$

$b = 0.1;$

```
c = 1;
F = @(t,y)[(1/b−a)*y(1)+y(3)+y(1)*y(2); −b*y(2)−y(1)*y(1); −y(1)−
c * y(3)];
[t,y] = ode45(F,[0,60],[.1,.2,.3]);
plot3(y(: ,1),y(: ,2),y(: ,3),'Color','b');
hold on
axis tight;
[t,y] = ode45(F,[60,120],[y(length(y(: ,1)),1),y(length(y(: ,2)),2),
y(length(y(: ,3)),3)]);
plot3(y(: ,1),y(: ,2),y(: ,3),'Color','r');
[t,y] = ode45(F,[120,180],[y(length(y(: ,1)),1),y(length(y(: ,2)),2),
y(length(y(: ,3)),3)]);
plot3(y(: ,1),y(: ,2),y(: ,3),'Color','g');
hold off
```

Project 3: A Damped Spring System

In this spring system, the spring has an object of mass m at the end. The damped spring can be modeled with the differential equation

$$m\frac{d^2x}{dt^2} + b\frac{dx}{dt} + kx = 0$$

where $k > 0$ represents the spring constant and the second term is the dampening term in the system.

a. Convert the equation to a system of first-order linear equations.

b. Determine the eigenvalues of the matrix associated with the system in part a. and use these values to find a general solution for the damped spring system.

c. Choose values for b, k, and m such that $b^2 - 4km > 0$ and explore the graph of the solution. Explain the behavior of the spring based on the graph.

d. Choose values for b, k, and m such that $b^2 - 4km = 0$ and explore the graph of the solution. Explain the behavior of the spring based on the graph.

e. Choose values for b, k, and m such that $b^2 - 4km < 0$ and explore the graph of the solution. Explain the behavior of the spring based on the graph.

f. Set amplitude=0 and explore different values for the mass, m, spring constant, k, and damping constant, b, in www.mathworks.com/matlabcentral/fileexchange/64747-forced-oscillator-with-dampening.

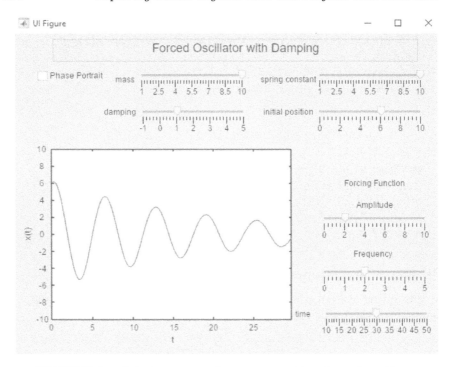

FIGURE 6.7: Solution curves for systems with a forced oscillator

Be sure to look both at the phase portrait and position graph so you can compare the results to those found in parts c through e.

Project 4: Romeo and Juliet

Researchers have studied how to model the romance between Romeo and Juliet with a coupled system of differential equations. The main question in this study is how will this romance change throughout time. The two variables in this study are $r(t)$, which is the love/hate of Romeo toward Juliet at time t and $j(t)$, which is the love/hate of Juliet toward Romeo at time t.

Note that if $j(t) > 0$ then Juliet loves Romeo at time t, if $j(t) = 0$ then Juliet's feelings toward Romeo are neutral at time t, and if $j(t) < 0$ then Juliet hates Romeo at time t.

Romeo's and Juliet's feelings for each other depend upon their partner's feelings and thus in the differential equation model, you will find interaction terms with interaction constants, p_1 and p_2. In addition, the rate at which Juliet's love is changing is dependent on the current amount of love that she possesses for Romeo. The rate at which Romeo's love for Juliet changes is also dependent on his current feelings. Producing the following model with the relationship between Romeo and Juliet,

$$j' = c_1 j + p_1 r,$$
$$r' = c_2 r + p_2 j.$$

a. If $c_1 = .5$, $c_2 = .5$, $p_1 = -.5$ and $p_2 = .6$, find the eigenvalues of the Jacobian matrix and determine the type of equilibrium point(s) that is present in the system. With an initial condition of $j(0) = 1, r(0) = 1$, interpret what will happen to Romeo and Juliet's relationship in the long run.

b. If $c_1 = -.5$, $c_2 = .5$, $p_1 = -.5$ and $p_2 = .6$, find the eigenvalues of the Jacobian matrix and determine the type of equilibrium point(s) that is present in the system. With an initial condition of $j(0) = 1, r(0) = 1$, interpret what will happen to Romeo and Juliet's relationship in the long run.

To visualize what is happening in part b. type:

```
c1 = -.5;
c2 = .5;
p1 = -.5;
p2 = .6;
F = @(t,y)[c1 * y(1) + p1 * y(2); c2 * y(2) + p2 * y(1)];
[t,y] = ode45(F,[0,50],[1,1]);
plot(y(: ,1),y(: ,2));
xlabel('Juliet');
ylabel('Romeo');
```

c. Explore the parameters c_1, c_2, p_1 and p_2 and initial conditions and determine values which will allow Romeo and Juliet's love to live forever.

Project 5: Modeling Epidemics

Using differential equations to model epidemics has been ongoing since the 1920s. The model that we will work with in this project is a stochastic differential equation model, predicting the probability of a behavior, and was proposed in 1964 by Bailey as a simple epidemic model.

$$\frac{dp_j}{dt} = (j+1)(n-j)p_{j+1}(t) - j(n-j+1)p_j(t), \text{ when } 0 \leq j \leq n-1,$$

$$\frac{dp_j}{dt} = -np_n(t), \text{ when } j = n,$$

where n is the total size of the population and p_j is the probability that there are j susceptible members of the community still unaffected by the epidemic.

a. If we write the system as $x' = Ax$, find A in terms of the above system.

b. If $n = 5$, determine the eigenvalues of A and their corresponding eigenvectors.

c. Find the Jordan canonical form, J, of A from part b.

d. Again using the matrix A from part b., find the matrix S where $S.J.S^{-1}$. How are the eigenvalues from part b. related to the columns of the matrix S?

e. Use the Jordan canonical form of A from part c. to determine a solution to the system of differential equations with initial condition $p_5(0) = 1$. To further explore the solution curves to this simple epidemic model see **www.mathworks.com/matlabcentral/fileexchange/64768-simple-epidemic-model**.

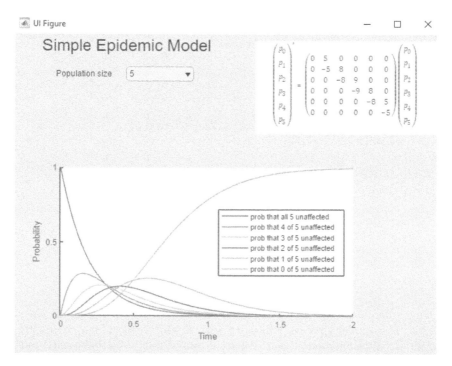

FIGURE 6.8: Solution curves for Bailey's simple epidemic model

MATLAB Demonstrations and References

MATLAB Demonstrations by Crista Arangala

All of the following MATLAB demonstrations are posted on the MATLAB Community File Exchange.

1. *Matrix Multiplication App*, https://www.mathworks.com/matlabcentral/fileexchange/63993-matrix-multiplication-app

2. *Permutations App*, http://www.mathworks.com/matlabcentral/fileexchange/64083-permutations-app

3. *Signed Determinant App*, https://www.mathworks.com/matlabcentral/fileexchange/64127-signed-determinant-app

4. 3×3 *Determinant App*, https://www.mathworks.com/matlabcentral/fileexchange/64140-3x3determinant-app

5. *Counting Paths of Nim App*, https://www.mathworks.com/matlabcentral/fileexchange/64175-counting-paths-of-nim-app

6. *Inverse and Nullspaces in Gf(p)*, https://www.mathworks.com/matlabcentral/fileexchange/65139-inverse-and-nullspaces-in-gf-p

7. *Hill Cipher App*,https://www.mathworks.com/matlabcentral/fileexchange/63769-hill-cipher-app

8. *Transforming the Dog*, https://www.mathworks.com/matlabcentral/fileexchange/64916-transforming-the-dog

9. *Transforming the Dog with Rotation*, https://www.mathworks.com/matlabcentral/fileexchange/64917-transforming-the-dog-with-rotation

10. *Transforming the Dog with a Composition of Linear Transformations*, https://www.mathworks.com/matlabcentral/fileexchange/66107-transforming-the-dog-with-a-composition-of-linear-transformations

11. *Sum of Two Vectors*, `https://www.mathworks.com/matlabcentral/fileexchange/64926-sum-of-two-vectors`

12. *Triangle Inequality with Functions*, `https://www.mathworks.com/matlabcentral/fileexchange/64935-triangle-inequality-with-functions`

13. *Cauchy–Schwarz for Vectors*, `https://www.mathworks.com/matlabcentral/fileexchange/64939-cauchy-schwarz-for-vectors`

14. *Cauchy–Schwarz for Integrals*, `https://www.mathworks.com/matlabcentral/fileexchange/64954-cauchy-schwarz-inequality-for-integrals`

15. *Change of Basis*, `https://www.mathworks.com/matlabcentral/fileexchange/64955-change-of-basis`

16. *Least Squares Linear Regression*, `https://www.mathworks.com/matlabcentral/fileexchange/64960-least-square-linear-regression`

17. *Conic Sections*, `https://www.mathworks.com/matlabcentral/fileexchange/64976-conic-sections`

18. *Multi-state Lights Out*, `https://www.mathworks.com/matlabcentral/fileexchange/65109-multistate-lights-out`

19. *Orthogonal Grids*, `https://www.mathworks.com/matlabcentral/fileexchange/65197-orthogonal-grids`

20. *Singular Values*, `https://www.mathworks.com/matlabcentral/fileexchange/65264-singular-values`

21. *Homogeneous Systems of Coupled Linear Differential Equations*, `https://www.mathworks.com/matlabcentral/fileexchange/64494-homogeneous-systems-of-coupled-linear-differential-equations`

22. *Visualizing the Solution of Two Linear Differential Equations*, `https://www.mathworks.com/matlabcentral/fileexchange/64580-visualizing-the-solutions-of-two-linear-differential-equations`

23. *Predator-Prey Model,*
    https://www.mathworks.com/matlabcentral/fileexchange/64676
    -predator-prey-system

24. *Forced Oscillator with Damping,* www.mathworks.com/matlabcentral/
    fileexchange/64747-forced-oscillator-with-dampening

25. *A Simple Epidemic Model,*
    https://www.mathworks.com/matlabcentral/fileexchange/64768
    -simple-epidemic-model

References

1. [C. Arangala et al. 2014], J. T. Lee and C. Borden, "Seriation algorithms for determining the evolution of The Star Husband Tale," *Involve*, 7:1 (2014), pp. 1-14.

2. [C. Arangala et al. 2010], J. T. Lee and B. Yoho, "Turning Lights Out," *UMAP/ILAP/BioMath Modules 2010: Tools for Teaching*, edited by Paul J. Campbell. Bedford, MA: COMAP, Inc., pp. 1-26.

3. [Atkins et al. 1999], J. E. Atkins, E. G. Boman, and B. Hendrickson, "A spectral algorithm for seriation and the consecutive ones problem," *SIAM J. Comput.* 28:1 (1999), pp. 297-310.

4. [D. Austin, 2013], "We recommend a singular value decomposition," A Feature Article by AMS, http://www.ams.org/samplings/feature-column/fcarc-svd, viewed December 12, 2013.

5. [N.T.J. Bailey, 1950], "A simple stochastic epidemic," *Biometrika*, Vol. 37, No. 3/4, pp. 193-202.

6. [E. Brigham, 1988], *Fast Fourier Transform and Its Applications*, Prentice Hall, Upper Saddle River, NJ, 1988.

7. [G. Cai and J. Huang, 2007], "A new finance chaotic attractor," *International Journal of Nonlinear Science*, Vol. 3, No. 3, pp. 213-220.

8. [P. Cameron], *"The Encyclopedia of Design Theory,"* http://www.designtheory.org/library/encyc/topics/had.pdf, viewed December 17, 2013.

9. [D. Cardona and B. Tuckfield, 2011], "The Jordan Canonical Form for a class of zero-one matrices," *Linear Algebra and Its Applications*, Vol. 235 (11), pp. 2942-2954.

10. [International Monetary Fund], *World Economic Outlook Database*, http://www.imf.org/external/pubs/ft/weo/2013/01/weodata/index.aspx, viewed December 20, 2013.

11. [J. Gao and J. Zhung, 2005], "Clustering SVD strategies in latent semantics indexing," *Information Processing and Management* 21, pp. 1051-1063.

12. [J. Gentle, 1998], *Numerical Linear Algebra with Applications in Statistics*, Springer, New York, NY, 1998.

13. [L. P. Gilbert and A. M. Johnson, 1980], "An application of the Jordan Canonical Form to the Epidemic Problem," *Journal of Applied Probability*, Vol. 17, No. 2, pp. 313-323.

14. [D. Halperin, 1994], "Musical chronology by Seriation," *Computers and the Humanities*, Vol. 28, No. 1, pp. 13-18.

15. [A. Hedayat and W. D. Wallis, 1978], "Hadamard matrices and their applications," *The Annals of Statistics*, Vol. 6, No. 6, pp. 1184-1238.

16. [K. Bryan and T. Leise, 2006], The "$25,000,000,000 Eigenvector," in the education section of *SIAM Review*, August 2006.

17. [J. P. Keener, 1993], "The Perron-Frobenius Theorem and the ranking of football teams," *SIAM Review*, Vol. 35, No. 1. (Mar., 1993), pp. 80-93.

18. The Love Affair of Romeo and Juliet, http://www.math.ualberta.ca/ ~ devries/crystal/ContinuousRJ /introduction.html, viewed December 22, 2013.

19. [I. Marritz, 2013] "Can Dunkin' Donuts really turn its palm oil green?," *NPR*, March 2013, viewed December 11, 2013. `http://www.npr.org/blogs/thesalt/2013/03/12/174140241/ can-dunkin-donuts-really-turn-its-palm-oil-green`.

20. [P. Oliver and C. Shakiban, 2006], *Applied Linear Algebra*, Prentice Hall, Upper Saddle River, NJ, 2006.

21. [One World Nations Online], Map of Ghana, http://www.nationsonline.org/oneworld/map/ghana_map.htm, viewed December 10, 2013.

22. [Rainforest Action Network], "Truth and consequences: Palm oil plantations push unique orangutan population to brink of extinction," http://www.npr.org/blogs/thesalt/2013/03/12/ 174140241/can-dunkin-donuts-really-turn-its-palm-oil-green, viewed December 11, 2013.

23. [K. R. Rao and P. C. Yip, 2001], *The Transform and Data Compression Handbook*, CRC Press, Boca Raton, FL, 2001.

24. [L. Shiau, 2006], "An application of vector space theory in data transmission," *The SIGCSE Bulletin*. 38. No 2, pp. 33-36.

25. [A. Shuchat, 1984], "Matrix and network models in archaeology," *Mathematics Magazine.* 57. No 1, pp. 3-14.

26. The University of North Carolina Chemistry Department, *Balancing Equations Using Matrices*, http://www.learnnc.org/lp/editions/chemistry-algebra/7032, viewed December 9, 2013.

27. Figure 6.3, http://www.scholarpedia.org/article/File:Equilibrium_figure_summary_2d.gif

Index